David Sharp

Fauna Hawaiiensis - Or the Zoology of the Sandwich (Hawaiian) Isles

Volume III, Part IV

David Sharp

Fauna Hawaiiensis - Or the Zoology of the Sandwich (Hawaiian) Isles
Volume III, Part IV

ISBN/EAN: 9783337272616

Printed in Europe, USA, Canada, Australia, Japan

Cover: Foto ©berggeist007 / pixelio.de

More available books at **www.hansebooks.com**

FAUNA HAWAIIENSIS

OR THE

ZOOLOGY OF THE SANDWICH (HAWAIIAN) ISLES:

Being Results of the Explorations instituted by the Joint Committee
appointed by

THE ROYAL SOCIETY OF LONDON FOR PROMOTING NATURAL KNOWLEDGE

AND THE BRITISH ASSOCIATION FOR THE ADVANCEMENT OF SCIENCE

And carried on with the assistance of those Bodies and of the Trustees of

THE BERNICE PAUAHI BISHOP MUSEUM AT HONOLULU.

EDITED BY

DAVID SHARP, M.B., M.A., F.R.S.

SECRETARY OF THE COMMITTEE.

VOLUME II. PART IV.

MOLLUSCA BY E. R. SYKES: *EARTHWORMS* BY F. E. BEDDARD:
ENTOZOA BY A. E. SHIPLEY.

Pages 271—441; Plate XI, coloured; Plates XII, XIII, XIV, uncoloured.

CAMBRIDGE:
AT THE UNIVERSITY PRESS.
1900

May 19, 1900.

MOLLUSCA
By E. R. SYKES, F.Z.S.

EARTHWORMS
By F. E. BEDDARD, F.R.S.

ENTOZOA
By A. E. SHIPLEY, M.A.

MOLLUSCA.

By E. R. Sykes,

WITH INTERCALATIONS ON ANATOMY

By Lt.-Col. Godwin-Austen.

Contents. § 1. *General remarks*, p. 271 ; § 2, *Systematic account*, p. 275 ; § 3. *Biblio-graphic list*, p. 400 ; § 4. *Alphabetical list of names placed as synonyms*, p. 407 ; § 5. *Alphabetical list of unidentified or erroneously recorded names*, p. 412.

THE material upon which this study of the land and fresh-water Mollusca is based, in addition to the collection formed by Mr Perkins, consists in a great measure of the collection in the British Museum (Natural History), where the bulk of Newcomb's and Pfeiffer's type-specimens are to be found. Thanks to the kindness of Prof. A. Hyatt, an examination has been made of the type-specimens of the species described by Gulick, of the genus *Leptachatina*, and now preserved at Boston. Mr D. D. Baldwin, of Maui, has also very kindly sent over a number of specimens and Mons. Ancey has lent the types of some species described by him. A collection formed in the Islands by Mr Hutchison has also been placed in my hands for examination by Mr Fulton.

Lt.-Col. H. H. Godwin-Austen, F.R.S., has enabled me to add to the interest of this work in a great degree, by very kindly dissecting some of the species and permitting me to incorporate here the results ; it is hoped that he may be able to give a further account of the anatomy in the *Achatinellidae*.

To Mr Edgar A. Smith, I desire, in conclusion, to express my most grateful thanks for his unfailing help and courtesy.

§ 1. General Remarks on the Mollusca.

To the student of the Mollusca, the Hawaiian Islands fauna is probably more familiar by name than that of many better known places, owing to the occurrence there of the well-known Achatinelloid group of forms.

Tables of the distribution of the fauna are given below, but a few general remarks here may be of interest.

The Limacidae yield nothing very peculiar or very striking and the few forms peculiar to the Islands may well have been developed from introduced European ancestors.

The Zonitidae are scattered over the Islands; all are peculiar, but they are nearly related to forms found in other islands of the Pacific: similar remarks apply to the Endodontidae, one group of which (*Pterodiscus*), however, appears to be peculiar.

The presence—and that strongly contested—of only a single indigenous species of the Helicidae again indicates affinities with Polynesia.

The Pupidae as a family, have a very wide geographical range, and hence no deductions can be drawn from their presence; it should be noted that here—if the identification be correct—the fauna includes a species not peculiar to the Islands.

With reference to the Achatinellidae it may at once be noted that several divisions of the family may be made. First, the brightly coloured forms which fall into the genus *Achatinella* proper and which are replaced in the Southern Pacific Islands by the genus *Partula*. The metropolis of distribution of all these forms seems to be Oahu, save in the case of the subgenus *Partulina* when Maui and Molokai appear to divide the honour. No species has been found on Kauai and only two on Hawaii at the other end of the group. Species have been described by authors upon coloration and band-formations; in my opinion numbers even of the 'species' here admitted will prove, when their anatomy is carefully investigated, to be varying forms of one common species. Consider, for example, such a shell as *Tachea nemoralis* dealt with in the same manner as the Hawaiian forms have been! Still, even when reductions are made, the fauna will remain remarkable for its numerical strength in species.

Secondly, passing through *Perdicella* and *Newcombia*, confined to the islands of Molokai and Maui, we come to the second great division, typified by *Leptachatina* and *Amastra*. Here, while the metropolis again seems to be Oahu, Kauai, the oldest island geologically considered, ranks well with the rest.

Thirdly, passing through the interesting and recently described *Thaanumia* of Oahu, we come to *Carelia*, which is confined to Kauai save for one subfossil species on the Island of Niihau (the only mollusc on that island).

Fourthly, we have the little group of *Auriculella* and *Frickella*, which leave the impression that they are linking forms between *Achatinella* and *Tornatellina*, and, again, belong in the main to Oahu. It should be borne in mind, as illustrating the peculiarity of the fauna, that only about half a dozen out of, approximately, 330 species of Achatinellidae are found on more than one island, and indeed some of these may be due to errors of identification. In our present state of knowledge a faunal list is largely influenced by the 'personal equation' of the writer.

From the residue of the fauna but little is to be learnt; the development of *Succinea* appears abnormal and further research will probably reduce the so-called 'species' of this group.

The following general conclusions may, however, be drawn :

1. The Molluscan fauna is nearly related to that of the Polynesian islands, and shows hardly any trace of continental influence, Asiatic or American.

2. The species are nearly always confined to one island ; but it is very doubtful if, as has been stated, "each valley has its peculiar species."

3. When the genera found are confined to the islands, the majority of living species usually occur on Oahu.

I give below tables of distribution ; but, owing to the fauna being so restricted in distribution, have not added percentages of peculiar species.

(1) Families Limacidae, Zonitidae, Endodontidae, Helicidae, Pupidae.

Species peculiar to one Island.

	Limacidae.	Zonitidae.	Endodontidae.	Helicidae.	Pupidae.	Total.
Kauai		3	8			11
Oahu		7	3		4	14
Molokai		1	1			2
Lanai	1	2				3
Maui	1	5	2			8
Hawaii	1	3	1	1	2	8

Species occurring in more than one Island.

Limacidae. Two species (*L. gagates* and *Agriolimax laevis*) are found elsewhere, and may have been introduced. One species is common to Kauai and Maui, one to Maui and Hawaii, and one to Kauai, Oahu, and Maui.

Zonitidae. One species common to Kauai, Oahu, and Maui ; one to Oahu, Molokai, and Lanai ; and one, respectively, to Kauai and Maui, Maui and Oahu.

Philomycidae. One found in Kauai, Oahu, and Hawaii, and one in Oahu only ; these species, however, are not peculiar to the Hawaiian fauna.

Endodontidae. Two species are common to Kauai and Oahu ; one, respectively, to Kauai and Lanai, Maui and Oahu, Lanai and Oahu, Lanai and Molokai ; while three are of uncertain habitat.

Helicidae. One species—introduced—in Kauai and Oahu.

Pupidae. One in Kauai, Oahu, and Hawaii ; one, respectively, in Kauai and Oahu, Oahu and Hawaii : further, a single species is found outside the Islands.

(2) Achatinellidae. In view of their interesting characters I have here dealt with the distribution by genera.

Species occurring in only one Island.

	ACHATINELLA group.											
	Achatinella s.l.	*Bulimella*	*Partulina*	*Achatinellastrum*	*Perdicella*	*Newcombia*	*Amastra*	*Leptachatina*	*Thaanumia*	*Carelia*	*Auriculella and Frickella*	*Total.*
Kauai							7	15		8	1	31
Oahu	17	19	1	29			40	33	1		10	150
Molokai			9	2	3	8	15	3			2	42
Lanai			3				12	5				20
Maui			16	3	5	1	17	4			1	47
Hawaii			2				3	7			1	13

The only forms of Achatinellidae found on more than one island are in the genera *Leptachatina* and *Auriculella*; in the former one species is said to be found on Kauai and Oahu, and two on Maui and Oahu; in the latter similar notes occur with regard to Oahu and Maui, Molokai and Lanai, Maui and Molokai, and (doubtfully) Oahu and Hawaii.

Further a single subfossil species of *Carelia* is recorded from Niihau; and the following are of uncertain habitat: *Bulimella* 1, *Partulina* 3, *Achatinellastrum* 1, *Amastra* 7, *Leptachatina* 6, and *Auriculella* 2.

(3) The residue of the fauna.

Species occurring in only one Island.

	Tornatellinae	STROBILIDAE	*Succinea*	LIMNAEIDAE	*Melania*	*Palaudestrina*	*Helicina*	*Neritina*	*Total.*
Kauai	3		3	2					8
Oahu	1		2	3	1	1	3		11
Molokai									0
Lanai	1		1						2
Maui	3		4	1	1				9
Hawaii	2	1	12	1					16

Species occurring in more than one Island.

Tornatellina. One species, said to be found in Oahu, occurs in the Tonga Islands. Two species are common to Kauai and Oahu, one to Hawaii and Oahu, and one to Kauai, Oahu, and Hawaii.

Stenogyridae. *Opeas junceus* is said to be found in all the Islands, and both this and *O. prestoni* (Hawaii) occur elsewhere. One species is of uncertain habitat.

Succinea. Two species in Oahu, Molokai, and Hawaii; one in Lanai, Oahu, and Maui; one in Kauai and Hawaii, one in Maui and Molokai.

Limnaeidae. One species in 'all the Islands'; one in Kauai and Oahu, one in Oahu and Maui; three of uncertain habitat.

Melania. One common to Kauai, Oahu, Maui, and Molokai; one to Kauai and Oahu; one to Kauai and Molokai; one of uncertain habitat.

Helicina. One common to Kauai, Oahu, Lanai, and Molokai; one to Maui and Lanai.

Neritina. One common to Maui, Oahu, and Hawaii; two of uncertain habitat. Two said to be found in 'all the Islands.'

These three tables show that Kauai has 50 species peculiar to it, Oahu 175, Molokai 44, Lanai 25, Maui 64, Hawaii 37.

§ 2. Systematic account of the fauna.

Fam. LIMACIDAE.

Amalia Moquin-Tandon.

Amalia M.-T. Hist. Moll. France, 1855, II. p. 19 [first species *Limax gagates*, Drap.].

Milax Gray, Cat. Pulm. Brit. Mus. 1855, p. 174 [has the same type; there are also, older names supposed to be identical, but founded on erroneous characters or improperly described].

While dealing with slugs it may be convenient to note that Semper has recorded a species stated to be very near *Limax tenellus* Nilsson; further Dr Cooper is said to have seen a species of *Janella* from these islands, but I have been unable to trace his note from the reference given (see Collinge, P. Malac. Soc. London, II. p. 50).

(1) *Amalia babori* Collinge.

Amalia babori Collinge, P. Malac. Soc. London, II. (1897), p. 294.

HAB. Maui, at 5000 ft., Haleakala.—Hawaii, 2000 to 4000 ft., Olaa to Kilauea (Perkins).

(2)　*Amalia gagates* Draparnaud.

Limax gagates Draparnaud, Tabl. Moll. France, 1801, p. 100 ; Hist. Moll. France, 1805, p. 122, pl. IX. figs. 1, 2.
Amalia gagates Drap., Collinge, P. Malac. Soc. London, II. p. 49.
HAB.　Maui (Perkins).

AGRIOLIMAX Mörch.

Agriolimax Mörch, J. Conchyl. XIII. (1865), p. 378.

As to the correct name for this genus, see Cockerell and Collinge, Conchologist, II. pp. 199, 200.

(1)　*Agriolimax bevenoti* Collinge.

Agriolimax bevenoti Collinge, P. Malac. Soc. London, II. (Nov. 1897), p. 295.
HAB.　Kauai, at 4000 ft.—Oahu, 2000 ft., Honolulu.—Maui, 5000 ft., Haleakala (Perkins).

(2)　*Agriolimax globosus* Collinge.

Agriolimax globosus Collinge, P. Malac. Soc. London, II. (April 1896), p. 47.
HAB.　Hawaii, Mauna Loa (Perkins).

(3)　*Agriolimax laevis* Müller.

Limax laevis Müller, Hist. Vermium, II. (1774), p. 1.
Agriolimax laevis Müller, Collinge, P. Malac. Soc. London, II. p. 295.
HAB.　Kauai, at 2000 ft., Lihue.—Maui, 5000 ft., Haleakala (Perkins).

(4)　*Agriolimax perkinsi* Collinge.

Agriolimax perkinsi Collinge, P. Malac. Soc. London, II. (April 1896), p. 47.
HAB.　Lanai, at 2000 ft. (Perkins).

(5)　*Agriolimax* (?) *sandwichiensis* Souleyet.

Limax sandwichiensis Souleyet, Voy. Bonite, Zool. II. (1852), p. 497, pl. XXVIII. figs. 8—11 [animal and shell].
HAB.　Hawaiian Islands (?).

It seems uncertain whether this be really Hawaiian, or even accurately represented ; see Collinge, P. Malac. Soc. London, II. p. 46.

Fam. ZONITIDAE.

GODWINIA, n. gen.

This new genus is proposed for the *Vitrina caperata* of Gould, which has, of recent years, usually been placed in *Helicarion*; it will be seen from the valuable anatomical notes of Lt.-Col. Godwin-Austen that there are differences which separate the species from that genus. Probably the *Vitrina tenella* of Gould also belongs here. The types of *Helicarion* Férussac (Tabl. Moll. 1821, pp. xxxi, 24) appear to have been the Australian forms *freycineti* and *cuvieri*.

(1) *Godwinia caperata* Gould.

Vitrina caperata Gould, P. Boston Soc. 11. (1847), p. 181; U. S. Explor. Exped. *Moll.* 1856, pl. 1. fig. 9.

Helix newcombi Pfeiffer, P. Zool. Soc. London, 1854 (Jan. 1855). p. 51; Reeve, Conch. Icon. *Helix*, pl. CLXXXIX. fig. 1321.

Plate XII. figs. 6—12.

HAB. Kauai (Gould, Perkins).—Oahu (Pfeiffer).

Very possibly the habitat of 'Oahu' is a mistake.

"The animal is dark, with a rather broad pale pallial margin; foot with a well defined central area beneath: the specimen was so much contracted that the mucous gland could not be decisively made out: from analogy, however, one should be present. There are no shell-lobes, the mantle-edge is curved and well defined. The right dorsal lobe is small, and the left lobe is long, narrow, and continuous.

"The visceral sac has three coils. The buccal mass has a strong, broad, muscle on the lower posterior side; the oesophagus is short, leading into a very capacious stomach; the salivary gland is in one compact, rounded mass. Jaw solid, dark sienna in colour, with a very straight cutting edge; odontophore long and narrow, with a few large median teeth; at first sight these centrals appear to be simple and straight-sided in form, and they are very nearly so, but closer examination shows that the centre and adjoining teeth have very small notches on the outer side; these are not cusps. The laterals are all curved and aculeate. The dental formula is:

$$18—5—1—5—18$$
$$23—1—23$$

Unfortunately the generative organs were not seen by me, all this portion being lost during dissection, as will sometimes occur in these small species.

" It will be seen from the above characters that this species cannot be placed in the Helicarionidae—the absence of shell-lobes forbids this. Aculeate laterals are hardly ever met with even in the genera of Zonitidae possessing shell-lobes; I can only recall one species, *Macrochlamys castaneolabiata*. The solid jaw, divided foot, and, in all probability, the presence of a mucous gland place it in the Zonitidae. In so many points is it distinct from any of the Indian and Malayan forms that I am acquainted with that I the more regret that the generative organs have still to be made out" (H. H. Godwin-Austen).

(2)   *Godwinia* (?) *tenella* Gould.

Vitrina tenella Gould, P. Boston Soc. II. 1847. p. 181; U. S. Explor. Exped. *Moll.* 1856, pl. 1. fig. 10.

HAB.   Kauai (Gould).—Maui, Haleakala, 5000—9000 ft. (Perkins).

The specimens found by Mr Perkins appear to be identical with Gould's species, which, so far as I can trace, has not been rediscovered on Kauai. In fresh specimens the lip is margined with black.

VITREA Fitzinger.

Vitrea Fitz., Beitr. Landeskund. Oesterr. III. p. 99.

Fitzinger's type, as I understand him, was *diaphana* Studer.

Until the anatomy of these Hawaiian species is known, I can suggest no better reference than to the present genus.

(1)   *Vitrea lanaiensis* Sykes.

Vitrea (?) *lanaiensis* Sykes, P. Malac. Soc. London, II. (1897), p. 298.

Plate XI. figs. 43, 44.

HAB.   Lanai, mountains behind Koele (Perkins).

(2)   *Vitrea molokaiensis* Sykes.

Vitrea (?) *molokaiensis* Sykes, P. Malac. Soc. London, II. (1897), p. 298.

Plate XI. figs. 45, 46.

HAB.   Molokai, forest above Pelekunu (Perkins).

(3) *Vitrea panxillus* Gould.

Helix pusillus Gould. P. Boston Soc. II. 1846, p. 171 [*non H. pusilla.* Lowe, 1831].

Helix panxillus Gould. U. S. Explor. Exped. *Moll.* p. 40, pl. III. fig. 46.

Hyalinia baldwini Ancey. Bull. Soc. Malac. France, VI. 1889, p. 192 ; Sykes, P. Malac. Soc. London, III. pl. XIII. figs. 1—3.

HAB. Maui (Gould); West part of Maui (Ancey); Haleakala, 5000 feet (Perkins). See, for a note on the synonymy, P. Malac. Soc. London, II. p. 298.

PSEUDOHYALINA Morse.

The original type was, I gather, *Helix exigua* Stimpson.

(1) *Pseudohyalina kauaiensis* Pfeiffer.

Helix kauaiensis Pfeiffer. P. Zool. Soc. London, 1854 [1855], p. 52 ; Reeve, Conch. Icon. *Helix.* sp. 1256.

HAB. Kauai (Pfeiffer).—Maui and Oahu (Baldwin).

I follow M. Ancey in the generic reference, as I do not know how, at present, the nomenclature may be bettered.

MICROCYSTINA Mörch.

Type *Nanina rinkii.* Mörch.

(1) *Microcystina* (?) *cryptoportica* Gould.

Helix cryptoportica Gould. P. Boston Soc. II. (1846), p. 20 ; U. S. Explor. Exped. *Moll.,* pl. V. fig. 72.

HAB. Oahu (Pease, Baldwin).

I place this here as the description states "columella valde intorta."

MICROCYSTIS Beck.

For a discussion as to the type, see P. Malac. Soc. London, II. pp. 130—2.

(1) *Microcystis chamissoi* Pfeiffer.

Helix chamissoi Pfeiffer, P. Zool. Soc. London, 1855, p. 91 ; Bland and Binney, Ann. Lyc. New York, x. p. 338, pl. xv. fig. 3 (jaw and radula ; copied in Ann. New York Ac. iii. pl. xvii. fig. O).

Hab. Kauai, Waioli and Haena (Baldwin) ; Makawele and Mountains above Waimea (Perkins).

Bland and Binney give "W. Maui" on the authority of Newcomb, but this seems very dubious ; Mons. Ancey gives Oahu, but I think this must be an error. The figures in the *Manual of Conchology* (Vol. ii. pl. xxxviii. figs. 74—6) are not good.

PHILONESIA, gen. nov.

Recently[1], I discussed the genus *Microcystis* Beck, and expressed the opinion that these small Zonitoid forms so characteristic of the Hawaiian Islands, and scattered over the Islands of the Central Pacific, could not be placed in that genus. I, further, referred them to *Macrochlamys*, stating that "whether our small forms are in accord with the typical group of this genus anatomically, remains to be proved ; but, conchologically, they only appear to differ in size."

Specimens of a form which I refer to the unfigured *Microcystis baldwini* Ancey, and which were collected by Mr Perkins, contained the animal, and Lt.-Col. Godwin-Austen has most kindly made an examination of it. His full report will be found on p. 281, but I may here summarize it by saying that this species does not belong to *Macrochlamys* at all, and the query I suggested has been answered. He points out its affinity to *Sitala* and *Kaliella* and here it is interesting to note that Mr Perkins found a species in the Hawaiian Islands that I have referred to the latter genus.

Under these circumstances, and as the shells are distinct by the conchological characters of the columella from both *Microcystina* and *Lamprocystis*—anatomically, also, from the former—I have ventured to create a new genus and propose to take *Microcystis baldwini* Ancey, as the type. Probably the bulk of the Hawaiian Zonitoid forms belong to this group.

(1) *Philonesia abeillei* Ancey.

Microcystis abeillei Ancey, Bull. Soc. Malac. France, vi. (1889), p. 199.

Hab. Molokai (Ancey) ; Mapulehu (Baldwin) ; wet forest above Pelekunu (Perkins).—Oahu, Waianae Mts. (Perkins).—Lanai (Perkins).

All the specimens are young, but I cannot sever them from this species.

[1] P. Malac. Soc. London, ii. p. 130.

(2) *Philonesia baldwini* Ancey.

Microcystis baldwini Ancey. Bull. Soc. Malac. France, vi. (1889), p. 204. Plate XII. figs. 1—5.

HAB. Oahu and west part of Maui (Ancey); Head of Panoa Valley, Nuuanu, and Honolulu Mts. (Perkins).

"The animal is brown; spotted and splashed with pure white (Plate XII. fig. 1 *a*) on the integument which covers the branchial chamber and visceral sac, these markings shew clearly through the transparent shell and give it a very pretty, mottled appearance. The extremity of the foot is truncated; with a mucous gland. In the specimen examined the foot (Plate XII. fig. 2) is very much contracted, but there is every indication that a small lobe overhangs the mucous gland. The foot, which is regularly segmented, has a central area (Plate XII. fig. 2 *a*); the pallial margin appears unusually broad, but this is deceptive and due to the extreme lateral contraction undergone; the two grooves above are similarly widened. The mantle edge has a well-developed, tongue-like, right shell lobe near the respiratory orifice, with an indistinct, narrow, left shell lobe. The right dorsal lobe is black and well developed, the left paler and moderately broad. Tentacles black.

"Plainly seen through the shell were four embryonic shells, lying one behind the other in the uterus, in various stages of development. The enveloping integument is transparent and so thin that the small shells, being comparatively heavy bodies, very readily break away, and the spermatophore adjacent was not made out.

"The odontophore has a formula of

$$30 : 9 : 1 : 9 : 30$$
$$39 : 1 : 39.$$

"The basal plates of the central teeth are quadrate in outline. The central tooth is tricuspid, the side cusps basal, blunt; the central point with convex sides. The median teeth have a blunt cusp only on the outer basal side, the ninth tooth is a narrower basal plate and is intermediate in form, the next eighteen being curved and bicuspid; the most interesting character is seen at this part of the row, for all the succeeding and outermost teeth are tricuspid, occasionally with even four points. The radula is remarkable for the similarity of the outermost teeth to those of *Kaliella barrakpurensis*[1]; those of *Sitala attegia* and *S. infula*[2] should also be compared, in which latter the pectiniform teeth are seen on the whole length of the row. The present shell shows an approach to *Kaliella* in a few of the outermost laterals, but it

---

[1] Land and F. W. Moll. India, i. pp. 19, 20, pl. v. fig. 11.
[2] Tom. cit. pl. viii. figs. 1 e & 2 e, after Stoliczka.

must be noted that the median teeth have a single outer cusp, while *Kaliella* has both outer and inner cusps: this latter characteristic is, however, not present in *Sitala*. *Kaliella* has few teeth in the row. *Sitala* many: 33 : 1 : 33, 153 : 1 : 153, respectively. A more important link with the genus *Sitala* is displayed by the presence of right and left shell lobes, which *Kaliella* does not possess: the close parallel lines of contraction across the right shell lobe shew that it has considerable extension in life. Stoliczka also mentions in *Sitala infula* the swollen uterus and the advanced state of development of the ova; pointing to similar embryonic stages in these molluscs. Yet another character is in common, namely, the absence of any amatorial organ. The male organ of the present species is also slightly different; I am unable, having only one specimen to dissect, to examine this in section.

"The jaw is very thin and delicate, and so colourless that its detection and extraction are very difficult. It has a well defined central projection on the cutting edge.

"The generative organs (Plate XII. figs. 3, 3*a*) cannot be described so fully as one would wish, owing to the expanded state of the uterus. The hermaphrodite duct and albumen gland were perfect: and the male portion thence complete. The prostate—as it is called by Semper, shewn in his figure of *Microcystis myops* as a loose fringe-like set of convolutions—appears in this species as a closely packed and thickened mass of oblong form, flattened on one side, where the oviduct would be lying attached if perfect. The vas deferens is given off at the anterior end. The penis is a thickened muscular tube, broad and bulbous below, tapering upwards to where the very short thickened retractor muscle is given off: the vas deferens at this point has three sharp convolutions; seen with transmitted light a short, sharp, 'kink' occurs in the bulbous portion near the generative aperture.

"The sculpture of the shell, magnified about thirty times, presents a very fine, regular, slightly wavy, longitudinally striated surface: this striation is strongest near the suture, becoming finer outwards. There are about 11 striae to ·003 inch. The most advanced embryonic shell consists of 2¼ whorls, the sculpture is well shewn on it.

"The point now to be solved is whether we are to retain this species in *Microcystis*. Mr Sykes regards[1] *M. ornatella* as the type of the genus; this was also the opinion of H. Nevill. Further Mr Sykes goes on to say 'Now these small Zonitoids [*i.e.* those of the Hawaiian Islands] hardly fit into the same genus as this species and therefore some other generic title is required for them.' The anatomy now described, shews, for many reasons, that the shell cannot be placed in *Macrochlamys* as Mr Sykes, guided by the shell characters, proposed. In my opinion it is undoubtedly close to *Kaliella*, still closer to *Sitala*, and yet there are sufficient differences in the generative organs to separate this Hawaiian form from both. If we take the shell alone into account, the sculpture presents one character, viz. fine, close longitudinal striation, not found in the Indian species of *Sitala*, in which the general surface is smooth, with spiral liration.

[1] P. Malac. Soc. London, II. 1896, p. 131.

The sculpture of *Kaliella* is finer and transverse to the whorl, so differs still more. It therefore may become necessary, if this shell be generally distinct from *M. ornatella*, to create a new genus.

"When we consider the immense area on the Equatorial belt over which *Kaliella*, *Sitala*, and this allied form are distributed, it appears that they fall naturally into a subfamily of their own which may be called the SITALINAE, Godwin-Austen, *nom. nov.*; one that is sufficiently distinct from the Durgellinae on the one hand, with which they are associated over a large portion of their range, and from the Macrochlaminae on the other, where the area of association is more restricted and the differences in the animal much greater." (H. H. Godwin-Austen.)

(3) *Philonesia cicercula* Gould.

Helix cicercula Gould, P. Boston Soc. II. (1846), p. 171; U.S. Explor. Exped. *Moll.* pl. v. fig. 73.

HAB. Hawaii (Gould); Kohala (Perkins).

var. *boettgeriana* Ancey.

Microcystis cicercula var. *boettgeriana* Ancey, Bull. Soc. Malac. France. VI. p. 206.

HAB. Hawaii, Kona (Ancey).

(4) *Philonesia exaequata* Gould.

Helix exaequata Gould, P. Boston Soc. II. (1846), p. 171; U.S. Explor. Exped. *Moll.* pl. v. fig. 61.
Helix disculus Pfeiffer, Zeitschr. für Malak. VII. 1851, p. 68 [*non* Deshayes].
Helix obtusangula Pfeiffer, *l. c.* p. 153.
Nanina discus Pfeiffer, Tryon, Man. Conch. Ser. II. Vol. II. p. 114.

HAB. Kauai (Gould, Perkins).

(5) *Philonesia hartmanni* Ancey.

Microcystis hartmanni Ancey, Bull. Soc. Malac. France. VI. (1889), p. 198.

HAB. Oahu (Ancey); Kalaikoa (Baldwin).

(6) *Philonesia indefinita* Ancey.

Microcystis indefinita Ancey, Bull. Soc. Malac. France, vi. (1889), p. 203.
Hab. Maui, east part (Ancey); Makawao (Baldwin).

(7) *Philonesia lymanniana* Ancey.

Microcystis lymanniana Ancey, Mem. Soc. Zool. France, vi. (1893), p. 329.
Hab. Oahu, Waialae (Ancey).

(8) *Philonesia oahuensis* Ancey.

Microcystis oahuensis Ancey, Bull. Soc. Malac. France, vi. (1889), p. 202.
Hab. Oahu (Ancey); Halemano (Perkins).
I refer, with some doubt, a single specimen found by Mr Perkins, to this unfigured species.

var. *depressiuscula* Ancey.

M. oahuensis var. *depressiuscula* Ancey, *l. c.* p. 203.
Hab. Oahu (Ancey).

(9) *Philonesia perlucens* Ancey.

Microcystis perlucens Ancey, Bull. Soc. Malac. France, vi. (1889), p. 207.
Hab. Maui, east part (Ancey).

(10) *Philonesia perkinsi* Sykes.

Macrochlamys perkinsi Sykes, P. Malac. Soc. London, ii. (1896), p. 126.
Plate XI. figs. 41, 42.
Hab. Lanai.—(?) Oahu, a single specimen (Perkins).

(11) *Philonesia platyla* Ancey.

Microcystis platyla Ancey, Bull. Soc. Malac. France, vi. (1889), p. 196; Sykes,
  P. Malac. Soc. London, iii. pl. xiii. figs. 13—15.
Hab. Oahu (Ancey); Waianae Mts. (Baldwin, Perkins).

(12) *Philonesia plicosa* Ancey.

Microcystis plicosa Ancey, Bull. Soc. Malac. France, VI. (1889), p. 200.
HAB. Oahu (Ancey); Palolo (Baldwin).

(13) *Philonesia sericans* Ancey.

Microcystis sericans Ancey, P. Malac. Soc. London, III. (1899), p. 268.
HAB. Hawaii, Olaa (Ancey).

(14) *Philonesia subrutila* Mighels.

Helix subrutila Mighels, P. Boston Soc. II. (1845). p. 19.
HAB. Oahu (Mighels, &c.).—Mr Baldwin gives Kauai, but I doubt this; the species is unknown to me.

(15) *Philonesia subtilissima* Gould.

Helix subtilissima Gould, P. Boston Soc. II. (1846), p. 177; U.S. Explor. Exped.
 Moll. pl. v. fig. 62.
Unknown to me: from the figure I am not certain of its generic position.
HAB. Maui (Gould).

(16) *Philonesia turgida* Ancey.

Microcystis turgida Ancey, Bull. Soc. Malac. France, VII. (1890), p. 339; Sykes.
 P. Malac. Soc. London, III. pl. XIII. figs. 5—7.
HAB. Maui (Ancey); Makawao (Baldwin); Mts. at 4000 ft. (Perkins).—A specimen found on Lanai by Mr Perkins, may belong to a variety.

Obs. The *Helix misella* of Férussac has been recorded with a query from the islands, but does not really belong to their fauna.

KALIELLA Blanford.

Type the group of *Helix barrakporensis* Pfr.

(1) *Kaliella konaensis* Sykes.

Kaliella konaensis Sykes, P. Malac. Soc. London. II. p. 299.
Plate XI. fig. 33.
A remarkable little shell which seems to fall between *Kaliella* and *Trochoconulus*.
Hab. Hawaii, Mt. Kona, at 3000 ft. (Perkins).

Fam. PHILOMYCIDAE.

Tebennophorus Binney.

Tebennophorus Binn., J. Boston Soc. IV. 1844. p. 171 (Type *Limax carolinensis*, Bosc).

(1) *Tebennophorus bilineatus* Benson.

Incilaria bilineata Benson, Ann. Nat. Hist. IX. 1842, p. 486.
Philomycus bilineatus Benson, Martens, Preuss. Exped. Ost-Asien, Mollusca, p. 16, pl. v. fig. 1.
Tebennophorus australis Bergh ?, Collinge, P. Malac. Soc. London, II. p. 50.
Tebennophorus striatus Hasselt, Collinge, *t. c.* p. 295.

Hab. Oahu, Mount Tantalus, Honolulu at 2000 ft.—Kauai, Lihue at 2000 ft.—
Hawaii, Olaa at 2000 ft. (Perkins).

See Collinge, J. Malac. VII. 1900, p. 80.

(2) *Tebennophorus striatus* Hasselt.

Meghimatium striatum Hasselt, Bull. Sci. Nat. Geol. III. 1824. p. 82.
Tebennophorus striatus Hasselt, Collinge, P. Malac. Soc. London, II. p. 50.
Hab. Oahu, Mount Tantalus (Perkins).

Fam. ENDODONTIDAE.

Endodonta Albers.

Endodonta Alb., Die Heliceen, 1850, p. 89 (first species *Helix lamellosa*, Fér.);
Op. cit. Ed. 2, 1860, p. 90 ("typus *Helix lamellosa*, Fér.").

(1) *Endodonta apiculata* Ancey.

Endodonta apiculata Ancey. Bull. Soc. Malac. France. VI. (1889). p. 189.

HAB.   Kauai. Dr Newcomb (Ancey).

(2) *Endodonta lamellosa* Férussac.

Helix lamellosa Férussac. Hist. Moll. I. p. 369. pl. LI. A. fig. 3 ; Quoy and
   Gaimard. Voy. Freycinet, Zool. p. 469. Pfeiffer, Conchylien-Cabinet. *Helix*,
   p. 197. pl. c. figs. 6—8.
Helix fricki Pfeiffer, P. Zool. Soc. London. 1858. p. 21. pl. XL. fig. 3.

According to Mörch (J. Conchyl. XIII. p. 395) this species "dépose ses œufs
dans l'ombilic."

The teeth or lamellæ seem to be variable ; some specimens shew traces of a second
tooth in the upper portion of the outer lip, thus having nine teeth in all.   Considerable
variation is also shewn in the relative proportions of height and breadth, and in the
width of the umbilicus.

HAB.   Oahu (Pease, Ancey); Waianae Mts. and Konahuanui (Baldwin); Mt.
Kaala,—Lanai Mts. behind Koele (Perkins).

(3) *Endodonta laminata* Pease.

Helix laminata Pease, Amer. J. Conch. II. (1866). p. 292.

According to the diagnosis this differs from the last by being spirally sculptured
as well as transversely ribbed, thereby becoming decussated.   The teeth appear to
be identical in number and position and I believe it will. eventually. only prove to be a
local race.

Tryon (Man. Conch. Ser. 2. III. p. 70) considered it to be a form of *E. cavernula*.
Hombr. and Jacq.. stating " I have before me two trays of shells named *Helix
laminata* Pease, from the 'Sandwich Is.,' one of them from the describer, which
undoubtedly represent the same species."   Since the two forms differ, from the
diagnoses. so widely in the armature. there must. I think. be some error.

HAB.   Kauai (Pease); Kahiliwi to Haena (Baldwin).

Sub-genus THAUMATODON Pilsbry.

(4) *Endodonta* (*Thaumatodon*) *contorta* Férussac.

Helix contorta Férussac, Hist. Moll. 1. p. 10, pl. LL. A. fig. 2.
Helix intercarinata Mighels, P. Boston Soc. 11. (1845). p. 18.

Specimens found by Mr Perkins, and which I refer to this species, appear to shew considerable variation and may be divided as follows :

A.   Six specimens, fairly typical in shape and size, but only one is furnished with five palatal teeth, the others having four.

B.   One specimen, darker in colouration, the colour markings being very distinct, palatal teeth five, these being remarkably incrassated.

C.   A long series (from Makaweli), larger, slightly more strongly sculptured and very variable in colouration, sometimes the dark brown colour predominating, at others a greenish yellow.   All appear to have four palatal teeth only.

HAB.   Oahu (various authors).—Kauai (Perkins).

(5) *Endodonta* (*Thaumatodon*) *hystricella* Pfeiffer.

Helix hystricella Pfeiffer, P. Zool. Soc. London, 1859. p. 25.

The original examples of this unfigured species, referred to as in Mus. Cuming, do not appear to be now in the British Museum.   Two specimens found by Mr Perkins agree well with Pfeiffer's diagnosis and dimensions ; they also accord in the number of teeth.

HAB.   Kauai (Pease).—Oahu, Kaala (Perkins).

(6) *Endodonta* (*Thaumatodon*) *nuda* Ancey.

Endodonta (*Thaumatodon*) *nuda* Ancey, P. Malac. Soc. London, 111. (1899). p. 268.
     pl. XII. fig. 1.

HAB.   Hawaii, Olaa (Ancey).

(7) *Endodonta* (*Thaumatodon*) *ringens* Sykes.

Endodonta (*Thaumatodon*) *ringens* Sykes, P. Malac. Soc. 11. (1896). p. 126.

Plate XI. figs. 39, 40.

In describing this species, I referred to it as having four teeth within the outer lip ; perhaps it would be more correct to say "one basal tooth and three within the outer lip." The ribs appear to be at varying distances apart. The Molokai specimens appear to belong to a large variety.

Hab. Lanai Mountains, behind Koele.—Molokai in wet forest above Pelekunu (Perkins).

(8) *Endodonta (Thaumatodon) rugata* Pease.

Helix rugata Pease, Amer. J. Conch. II. (1866), p. 291.

Hab. Maui (Pease).

Sub-genus Nesophila Pilsbry.

The following table may assist in separating the species of *Nesophila*.

A. Parietal lamellæ absent.—*E. capillata* Pease.

B. Parietal lamella single.—*E. decussatula* Pease; *E. elisae* Ancey; *E. jugosa* Mighels; *E. lanaiensis* Sykes; *E. stellula* Gould.

C. Parietal lamellæ two.—*E. binaria* Pfeiffer; *E. hystrix* (Mighels MS.) Pfeiffer; *E. paucicostata* Pease.

D. Parietal lamellæ several.—*E. baldwini* Ancey; *E. distans* Pease; *E. tiara* Mighels.

(9) *Endodonta (Nesophila) baldwini* Ancey.

Charopa baldwini Ancey. Bull. Soc. Malac. France, VI. (1889), p. 176.

Mons. Ancey also records a white variety.

Hab. Hawaiian Islands (Ancey).

(10) *Endodonta (Nesophila) binaria* Pfeiffer.

Helix binaria Pfeiffer, P. Zool. Soc. London, 1856, p. 33.

I am unable to trace the type of this species, which should have passed with Cuming's collection into the British Museum.

Hab. Kauai (Pease).

(11) *Endodonta (Nesophila) capillata* Pease.

Helix capillata Pease, Amer. J. Conch. II. (1866), p. 292.

Hab. Kauai (Pease).

(12)   *Endodonta (Nesophila) decussatula* Pease.

Helix decussatula Pease. Amer. J. Conch. II. (1866), p. 291.

HAB.   Molokai (Pease); Mountains at 4000 ft. (Perkins).

Mr Baldwin gives " Kauai " as the habitat, but, since he marks it as a species he has not seen, I think there is probably some error.

(13)   *Endodonta (Nesophila) distans* Pease.

Helix distans Pease. Amer. J. Conch. II. (1866), p. 290.

HAB.   Kauai (Pease).

(14)   *Endodonta (Nesophila) elisae* Ancey.

Pitys elisae Ancey, Bull. Soc. Malac. France, VI. (1889), p. 180.

Unknown to me.

HAB.   ? Hawaiian Islands (Ancey).

(15)   *Endodonta (Nesophila) hystrix* (Mighels MS.) Pfeiffer.

Helix hystrix Pfeiffer, Symb. Hist. Hel. III. p. 67 : Gould. U. S. Explor. Exped. *Moll.* pl. IV. fig. 52*.

Helix setigera Gould, P. Boston Soc. I. p. 174 [nec Sowerby].

It is, of course, not the *Helix hystrix* of Cox, an Australian species.

HAB.   Oahu (authors); Mount Kaala, Oahu (Perkins).

(16)   *Endodonta (Nesophila) jugosa* Mighels.

Helix jugosa Mighels, P. Boston Soc. II. (1845), p. 19.

Helix rubiginosa Gould, P. Boston Soc. II. (1846), p. 173; U. S. Explor. Exped. *Moll.* pl. IV. p. 49.

The two forms were first united by Pease[1], who stated that the synonymy was accepted by Gould; recently Mons. Ancey[2] has revived *E. rubiginosa* as a species, referring to it some shells from Oahu. In this state of conflict I have followed Pease, considering that he and Gould were in the best position to form an opinion.

HAB.   Kauai, Waioli to Kapaa (Baldwin); Kauai (Perkins, etc.).

[1] J. Conchyl. XIX. (1870), p. 95.

[2] Bull. Soc. Malac. France, VI. (1889), p. 179.

(17) *Endodonta (Nesophila) lanaiensis* Sykes.

Endodonta (Nesophila) lanaiensis Sykes, P. Malac. Soc. London, II. (1896), p. 127.
Plate XI. figs. 37, 38.

Specimens from Kauai which I refer to this species are strongly hispid in the young state, but with age the hairs appear to be rubbed off; none of those from Lanai are very young, and only traces of hairs can be seen. The species appears to be near *E. decussatula*, but it almost lacks decussation and is darker in colour; the interstices of the ribs are closely, finely, striate. Save for the presence of a parietal lamella, the Kauai specimens approach Pease's diagnosis of *E. capillata*.

HAB. Lanai Mountains, behind Koele.—Kauai. Makaweli, on *Dracaena* and *Cheirodendron* (Perkins).

(18) *Endodonta (Nesophila) paucicostata* Pease.

Helix paucicostata Pease. J. Conchyl. XVIII. (1870). p. 393.
Helix filocostata Pease, P. Zool. Soc. London, 1871, p. 454.
HAB. Kauai (Pease).

(19) *Endodonta (Nesophila) stellula* Gould.

Helix stellula Gould, P. Boston Soc. I. (1844), p. 174: U. S. Explor. Exped.
 Moll. pl. IV. fig. 52†.
HAB. Maui (Gould).

(20) *Endodonta (Nesophila) tiara* Mighels.

Helix tiara Mighels, P. Boston Soc. II. (1845), p. 19; Kuster, Conch.-Cab. *Helix.*
 pl. CXXV. figs. 9—11.

According to Mons. Ancey[1] this species possesses several parietal lamellæ; if so, the character has been omitted from the various diagnoses.

HAB. Kauai (various authors).

(21) *Endodonta (Nesophila)*. sp.

Two interesting little specimens with a depressed spire were found on Molokai by Mr Perkins, the exact habitat being " Forest above Pelekunu "; they have 4–4½ whorls, with two parietal lamellæ and no teeth within the other lip, but appear not to be adult.

HAB. Molokai.

[1] Bull. Soc. Malac. France, VI. p. 175.

PTERODISCUS Pilsbry.

Type. *P. wesleyi* Sykes.

(1) *Pterodiscus digonophorus* Ancey.

Patula digonophora Ancey, Bull. Soc. Malac. France. VI. (1889). p. 171 ; Sykes.
P. Malac. Soc. London, III. pl. XIII. figs. 9—11.

HAB. Oahu (Ancey); Waianae Mts. (Baldwin).

(2) *Pterodiscus petasus* Ancey.

Pterodiscus petasus Ancey, P. Malac. Soc. London. III. (1899). p. 268. pl. XII.
fig. 4.

HAB. Oahu. Waianae Mts. (Ancey).

(3) *Pterodiscus wesleyi* Sykes.

Endodonta (Pterodiscus) wesleyi Sykes, P. Malac. Soc. London, III. (1896). p. 127.
Endodonta (Pterodiscus) alata Pfeiffer, Pilsbry. Man. Conch. (2) IX. p. 36. pl. IV.
fig. 44 [nec *Helix alata*, Pfeiffer].

HAB. Hawaiian Islands.

The following two species, placed in this group by Mr Pilsbry[1]. with the habitat of
Hawaiian Islands, are unknown to me. They were originally described from " Islands
of the Central Pacific " by Pease : the first has been recorded from Tahiti, but never
again found there, and Mons. Ancey has suggested a Hawaiian origin : the second has
been localized as from (?) Lanai. They are *Helix prostrata* and *H. depressiformis*
(P. Zool. Soc. London, 1864, p. 670).

Fam. HELICIDAE.

PAPUINA von Martens.

Papuina Mart., Die Heliceen, Ed. 2, 1860. p. 166 (type *Helix litmus*, Lesson).

(1) *Papuina barnaclei* Smith.

Helix (Merope?) barnaclei Smith, Ann. Nat. Hist. (4) XX. p. 242.

I am informed that careful search has been made, in the neighbourhood indicated,
but that no trace of the species can be found. At present, bearing in mind on the one

---

[1] Man. Conch. (2) IX. p. 36.

hand the improbability of a species of *Papuina* occurring in Hawaii, and on the other the positive statement of the original collector, I can but include it, with this note of warning.

HAB. Hawaii, eight miles from Kailua (Smith).

EULOTA Hartmann.

Eulota Hart., Erd- und Susswasser Gasteropoden, p. 179 (type *Helix fruticum*. Müll.). The date usually given is 1842, but the title-page of the copy in the British Museum bears that of 1840.

(1) *Eulota similaris* Férussac.

Helix similaris Férussac, Prodrome, 1822, p. 47 (*nom. sol.*); Hist. Moll. i. p. 171. pl. XXV. B, figs. 1—4. XXVII. A, figs. 1—5.

A widely scattered species; presumably not indigenous.

HAB. Kauai (Pilsbry).—Oahu, Tantalus (Perkins).

The following have been described under the term *Helix* and recorded from the Islands.

Helix fornicata Gould. P. Boston Soc. II. (1846). p. 172.

Supposed to come from Kauai. Tryon notes[1]: "In the corrigenda to the Mollusca of the Wilkes Exploring Expedition, Dr Gould states that the only specimen was lost, and *H. tongana* Quoy, figured by the artist for this species."

Helix sandwichensis Pfeiffer, P. Zool. Soc. London, 1849, p. 128.

Appears to be the young of a South American *Systrophia.*

Helix exserta Pfeiffer, P. Zool. Soc. London, 1856, p. 32.

Only a fragment of the type remains; it has never been figured.

Fam. PUPIDAE.

PUPA Draparnaud (1801).

Pupa Drap., Tabl. Moll. France, pp. 32, 56 (first species *Turbo muscorum* L.); 1805, Hist. Moll. France. p. 59.

There appears to be a *Pupa* of Lamarck of even date (Syst. anim. sans Vert. p. 88) with *Turbo uva* as type; also, through the kindness of Mr Sherborn, I have

---

[1] Man. Conch. (2), III. p. 27.

examined the *Museum Boltenianum*, Ed. 1. 1798, and Bolten proposed *Pupa* (p. 110) for *Voluta flammea* and *V. solidula*. The first species belongs to *Actaeon* (1810) and the second is the type of *Solidula* (1807). If, therefore, Bolten's names are to be used, *Pupa* Drap. cannot stand unless it prove to be the same as *Pupa* Humphrey, 1797 (Mus. Calonnianum).

An excellent study of the Pupidae of Oceania has been written by Dr Boettger; the attention of subsequent authors does not appear to have been sufficiently directed to this paper.

(1) *Pupa acanthinula* Ancey.

Pupa acanthinula Ancey, Mem. Soc. Zool. France, v. (1892), p. 709.
HAB. Oahu, Makiki (Ancey).

(2) *Pupa admodesta* Mighels.

Pupa admodesta Mighels, P. Boston Soc. II. (1845), p. 19; Boettger, Conch.
    Mittheil. I. p. 58, pl. XII. fig. 15.
HAB. Oahu (Mighels).

(3) *Pupa bacca* Pease.

Pupa bacca Pease, P. Zool. Soc. London, 1871, p. 462.
HAB. Hawaii, Kalapana (Pease).

(4) *Pupa costata* Pease.

Pupa costata Pease, P. Zool. Soc. London, 1871, p. 462.
HAB. Hawaii (Pease).

(5) *Pupa lyonsiana* Ancey.

Pupa lyonsiana Ancey, Mem. Soc. Zool. France, v. (1892), p. 713.
HAB. Oahu, Punahou (Ancey).

(6) *Pupa lyrata* Gould.

Pupa lyrata Gould, P. Boston Soc. I. (1843), p. 139; Boettger, Conch. Mittheil.
    I. p. 61, pl. XII. fig. 17.
Pupa striatula Pease, P. Zool. Soc. London, 1871, p. 461.
Pupa magdalenae Ancey, Mem. Soc. Zool. France, v. (1892), p. 716.

I follow Dr Boettger in placing *P. striatula* with *P. lyrata*, and have added *P. magdalenae*, since the character of two parietal teeth, on which Mons. Ancey lays special stress, is found in some undoubted specimens of *P. lyrata*.

HAB. Hawaii (Pease).—Oahu (Baldwin) ; Palaina (Ancey).

(7) *Pupa mirabilis* Ancey.

Pupa mirabilis Ancey, Bull. Soc. Malac. France, VII. (1890). p. 339.
HAB. Oahu (Ancey).

(8) *Pupa newcombi* Pfeiffer.

Pupa newcombi Pfeiffer, P. Zool. Soc. London, 1852 [1854], p. 69 ; Kuster, Conch.-Cab. *Pupa*, pl. XX. figs. 23, 24.
Pupa costulosa Pease, P. Zool. Soc. London, 1871. p. 462.
HAB. Hawaii (Pease. Baldwin).—Oahu and Kauai (Baldwin) [sed quaere, E.R.S.].

var. *seminulum* Boettger.

Pupa newcombi var. *seminulum* Boettger, Conch. Mittheil. I. p. 58. pl. XII. fig. 14.
HAB. Probably Hawaii (Boettger).

(9) *Pupa pediculus* Shuttl. var. *nacca* Gould.

Vertigo nacca Gould, P. Boston Soc. VIII. (1862). p. 280.
Dr Boettger has pointed out that this is only a form of *P. pediculus*, Shuttleworth (Bern. Mittheil. 1852. p. 296), and barely of varietal rank. The typical form appears not to be found in the Hawaiian Islands.

HAB. Hawaii (Gould).

(10) *Pupa perlonga* Pease.

Pupa perlonga Pease, P. Zool. Soc. London, 1871, p. 462 ; Boettger, Conch. Mittheil. I. p. 60, pl. XII. fig. 16.

Only two specimens, which I refer here with some hesitation, since the teeth are rather obscure.

HAB. Oahu (Pease).—Kauai, Makaweli (Perkins).

Fam. ACHATINELLIDAE.

The various species of this family form probably one of the most puzzling groups of land-shells known : numerous 'species' have been described, founded almost entirely on shell colouration or banding, and this, bearing in mind such protean forms as *Tachea nemoralis* or *Polymita picta*, is a course which often leads to endless trouble.   At present, such genera as *Achatinella* s. s. and *Achatinellastrum* are in utter confusion, and any attempt at a list of species simply reflects the general view of the writer and cannot be regarded as authoritative.

Of recent years some species have been described, based on shell-colouration combined with that of the mantle, but it should be remembered that some malacological characteristics are frequently as liable to variation as are conchological ones.   The severance of species by consideration of habitat is, again, not a necessarily certain guide, since one species of mollusc may range widely, while at different points of its range, local variations may occur.

The history of the Achatinellidae in literature commences with Dixon's description of *Turbo apexfulva* ; subsequently stray species were described by various authors, but no serious attempts were made at their study until about 1850 60, when Newcomb and Pfeiffer added greatly to our knowledge.   To Mr Gulick, in 1858, we owe large additions to the catalogue—so far as names go—but his views of species were somewhat too narrow, and he described many trifling variations as species, mainly on the ground of geographical range.   Later, Pease did useful work and gave a catalogue of the family. Dr Hartman, in 1888, listed the known forms in a similar way.   Of recent years Mr Baldwin has done the chief work on the group, and his very valuable catalogue has been of great assistance.

Numerous genera and sections have been described, almost all on purely conchological grounds, and so variable are the forms that linking species are easily found. Still they prove useful in the arrangement of species and therefore have been admitted in this work.

A list of them arranged in chronological order may be of use.

1828.   *Achatinella* Swainson, Q. J. Sci. Lit. and Arts. p. 81.

1847.   *Leptachatina* Gould, P. Boston Soc. 11. p. 201.

1854.   *Partulina* Pfeiffer, Malak. Blatt. 1. p. 114.

1854.   *Newcombia* Pfeiffer, *l. c.* p. 117.

1854.   *Bulimella* Pfeiffer, *l. c.* p. 119.

1854.   *Laminella* Pfeiffer, *l. c.* p. 126.

1854.   *Achatinellastrum* Pfeiffer, *l. c.* p. 133.

1854.   *Labiella* Pfeiffer, *l. c.* p. 142.

1855. *Carelia* H. & A. Adams, Gen. Rec. Moll. II. p. 132.

1855. *Amastra* H. & A. Adams, *l. c.* p. 137.

1855. *Auriculella* Pfeiffer, Malak. Blatt. II. p. 3.

1855. *Frickella* Pfeiffer, *l. c.* p. 3.

1860. *Apex* Martens, Die Heliceen, p. 248.

1870. *Eburnella* Pease, P. Zool. Soc. London, 1869, p. 647.

1870. *Perdicella* Pease, *l. c.* p. 649.

1875. *Cariuella* Pfeiffer, Novit. Conch. IV. p. 116.

1899. *Thaanumia* Ancey, P. Malac. Soc. London, III. p. 269.

1899. *Baldwinia* Ancey, *l. c.* p. 270.

Mr Pilsbry's remark, that the use of Férussac's term *Helicteres* "would open the door to an endless series of vagaries in nomenclature," appears to me to be thoroughly justified; for the converse view see Pease, P. Zool. Soc. London, 1862, p. 3. The *Achatinella* of Schlüter (1838) appears to be a mixture of *Ferussacia* and allied groups.

Our knowledge of the anatomy is due primarily to Bland and Binney, who pointed out that the Achatinellidae may be divided into two main groups based on the characters of the radula and jaw; also that *Carelia*, while it agrees in the radula with the *Leptachatina* and *Amastra* group, differs in the characters of the jaw. Heynemann has also written on the subject, and recently notes have been published by Messrs Gwatkin, Pilsbry and Suter. Mr Pilsbry has recently[1] stated with reference to *Partula* and *Achatinella* that "these forms have no relations with the Bulimulidae and Achatinidae, with which conchologists associate them, but lie at the base of the terrestrial pulmonate tree."

The classification here adopted is:

Genus *Achatinella*. Subgenn. *Achatinella* s. s. (Syn. *Apex*.) *Bulimella*. *Partulina*. (Syn. *Baldwinia*.) *Achatinellastrum*. (Syn. *Eburnella*.)

Genus *Perdicella*.

Genus *Newcombia*.

Genus *Amastra*. Subgenn. *Amastra* s. s. *Laminella*. *Amastrella*, n. subgen. *Kauaia*, n. n. [*Cariuella*. Pfr., nec Sby.]

Genus *Leptachatina*.

Genus *Thaanumia*.

Genus *Carelia*.

Genus *Auriculella*.

Genus *Frickella*.

[1] Ann. nat. Hist. IV. (1899). p. 156.

ACHATINELLA s. str.

Synonym. *Apex* von Martens.

Well has Dr Hartman remarked. that the species of this group "are involved in almost inextricable confusion." Our ignorance of the animal in most forms, combined with the fact that authors have described frequently from single specimens, or from series of two or three, entirely at present prevents one ascertaining the true specific limits.

In the one or two instances where I have attempted 'lumping,' the results are due to a long series of specimens in which I have been unable to arrive at a definite break between the one 'species' and the other.

The conclusions I have drawn are founded on Newcomb's and Pfeiffer's type specimens; specimens compared with Gulick's types: further a few of the types described by Mr Smith; and, finally, the specimens collected by Mr Perkins and a collection formed by Mr Hutchison.

All the species of this group are confined to the Island of Oahu.

(1)　*Achatinella albospira* Smith.

Apex albospira Smith, P. Zool. Soc. London, 1873. p. 77. pl. x. fig. 8.

Unknown to me; it has been united by Dr Hartman with *A. turgida* Newc. [= *A. cestus* Newc.].

HAB.　Oahu, Ewa.

(2)　*Achatinella apexfulva* Dixon.

Turbo apex fulva Dixon, Voyage round the world, 1789, p. 344, figd. on an unnumbered plate.

Turbo lugubris Chemnitz, Conch.-Cab. XI, 1795. p. 278. pl. CCIX. figs. 2059, 2060.

Cochlogena (Helicteres) lugubris Chemnitz, Férussac, Prodrome, p. 60, no. 431.

Monodonta seminigra Lamarck, Hist. Nat. Anim. sans Vert. VII. (1822). p. 37.

Achatinella pica Swainson, Zool. Ill. *Moll.* ser. II. pl. XCIX. fig. 1.

HAB.　Oahu. Kawailoa (Baldwin).

Dixon was thoroughly binomial, as a reference to his work will show: his figures are good, and the only point that can be made against the utilization of his name is that it appeared in three words. This, probably, was a printer's error, or a slip, as other names in the same work are proper, and suitable descriptions are given.

(3) *Achatinella apicata* (Newcomb MS.) Pfeiffer.

Achatinella apicata Pfeiffer, P. Zool. Soc. London, 1855 [1856], p. 210.
HAB. Oahu, Halemano (Perkins).

var. *polymorpha*, Gulick.

Apex polymorpha Gulick, P. Zool. Soc. London, 1873, p. 81, pl. x. fig. 5.

HAB. Oahu, Wahiawa, Kalaikoa, Ahonui (Gulick); Kawailoa Gulch, and above Ewa (Perkins).

var. *vespertina* Baldwin.

Achatinella (Apex) vespertina Baldwin, P. Ac. Philad. 1895, p. 219 [shell and animal], pl. x, fig. 14.

HAB. Oahu, Kawailoa (Baldwin); from a ridge between Waala and Kawailoa Gulches (Perkins).

var. *alba, var. n.*

Shell snow-white, save for the peristome being margined with lilac, similar tinting appearing on the columella plait, and inside the outer lip at its junction with the body-whorl of the shell.

HAB. Oahu, near head of Kawailoa Gulch (Perkins).

A fine series of this species. It appears to be distinct from *A. swiftii* Newc. [= *A. cestus* Newc. var.], to which Newcomb subsequently referred it, and the type of which I have examined. The shell is larger, different in form, and more polished, the ground colouring darker, and the banding not so conspicuous in the typical form: the protoconch is in general brown, and but rarely white. The prevailing tint is brown, sometimes becoming black with a white area below the suture, rarely altogether snow-white. A few, from near Kawailoa, are tinted with lilac on an ashy ground, occasionally having a chestnut sutural line; these form a passage to *A. vespertina* Baldwin, in which the lilac tint has changed to cream-colour. This latter is the only form in which the animal appears to have been noticed.

(4) *Achatinella bicolor* (Gulick) Pfeiffer.

Achatinella bicolor Pfeiffer, Mon. Hel. Viv. IV. 1859. p. 529.
Near, apparently, to *A. cookei* Baldwin.
HAB. Oahu, Lehui (Pfeiffer).

(5) *Achatinella cestus* Newcomb.

Achatinella cestus Newcomb, P. Zool. Soc. London, 1853 [1854]. p. 132. pl. XXII.
fig. 8.
Achatinella turgida Newcomb, *tom. cit.* p. 134. pl. XXII. figs. 10, 10 *a*.

HAB. Oahu, Palolo and Ewa (Newcomb); Ewa and Halemano (Perkins).

var. *swiftii* Newcomb.

Achatinella swiftii Newcomb, P. Zool. Soc. London, 1853 [1854]. p. 133.
pl. XXII. fig. 9; Ann. Lyc. New York, VI. p. 325 [animal].
Achatinella (Achatinellastrum) valida Pfeiffer, P. Zool. Soc. London, 1855. p. 6,
pl. XXX. fig. 24.
Apex flavidus Gulick, *op. cit.* 1873. p. 80, pl. X. fig. 1.
Apex tuberans Gulick, *tom. cit.* p. 81, pl. X. fig. 3.
Apex liliaceus Gulick, *tom. cit.* p. 79, pl. X. fig. 4.
Apex turbiniformis Gulick, *tom. cit.* p. 81, pl. X. fig. 7.

HAB. Oahu, Ewa (Newcomb); as *A. flavidus*, Kalaikoa and Ahonui (Gulick);
as *A. tuberans*, Kalaikoa, Ahonui, Wahiawa and Halemano (Gulick); as *A. turbiniformis*, Kalaikoa and Lehui (Gulick).

var. *forbesiana* Pfeiffer.

Bulimella forbesiana Pfeiffer, P. Zool. Soc. London, 1855. p. 3, pl. XXX. fig. 16.
Apex gulickii Smith, *loc. cit.* 1873, p. 78, pl. IX. fig. 19 [*non* 17].

HAB. Oahu as *A. gulickii*, Kalaikoa, Ahonui and Waialei (Smith).

The type form has the bands interrupted; in the variety *swiftii* they are continuous
and almost confluent, recalling a dwarf *A. lugubris*. From the variety *swiftii*, given a
fair number of specimens, we may pass by slight variations up to a nearly white shell
with hardly a trace of banding. One specimen is light yellow and almost unmarked
except the latter half of the last whorl, which is strongly banded with dark brown (Plate
XI. figs. 6. 7). The general brown colouring is more persistent in the variety than in
the typical form. The variety *forbesiana* is somewhat more elongate and slightly
differently banded and coloured.

(6) *Achatinella cookei* Baldwin.

Achatinella (Apex) cookei Baldwin, P. Ac. Philad. 1895. p. 220. pl. X. fig. 15
[shell and animal]; Suter, *t. c.* p. 239 [radula].

HAB. Oahu, Waiau (Baldwin).

(7)   *Achatinella decora* Férussac.

Cochlogena (Helicteres) decora Férussac, Prodrome, 1822, p. 60, no. 430.
Turbo lugubris sinistrorsus Chemnitz, Conch.-Cab. XI. 1795, p. 307, pl. CXIII.
    figs. 3014, 3015.
Achatinella perversa Swainson, Quart. Journ. Sci. &c. I. 1828, p. 84; Zool. Ill.
    1833, ser. II. pl. XCIX. fig. 2; Newcomb, Ann. Lyc. New York, VI. 1858,
    p. 309 [animal].
Achatinella concidens Gulick, Ann. Lyc. New York VI. 1858, p. 234, pl. VIII.
    fig. 54.
HAB.   Oahu, Halemano, Waimea.

var. *leucophaea* Gulick.

Apex leucophaeus Gulick, P. Zool. Soc. London, 1873, p. 82, pl. IX. fig. 16.
HAB.   Oahu, Waialei.

var. *neglecta* Smith.

Apex neglectus Smith, P. Zool. Soc. London, 1873, p. 78, pl. IX. fig. 22.
Apex coniformis Gulick, tom. cit. p. 81, pl. IX. fig. 17.
Hab.   Oahu, Wahiawa, Kalaikoa, Ahonui, Halemano (Smith, Gulick).

var. *versicolor* Gulick.

Apex versicolor Gulick, P. Zool. Soc. London, 1873, p. 80, pl. IX. fig. 18.
HAB.   Oahu, Ahonui, Kalaikoa.

var. *innotabilis* Smith.

Apex innotabilis Smith, P. Zool. Soc. London, 1873, p. 78, pl. IX. fig. 23.
HAB.   Oahu.

subspecies *mustelina* Mighels.

Achatinella mustelina Mighels, P. Boston Soc. II. 1845, p. 21; Reeve, Conch. Icon.
    Achatinella, pl. III. fig. 21; Newcomb, Ann. Lyc. New York, VI. p. 313
    [animal].
HAB.   Oahu, Waianae, Kaala.

var. *leucorraphe* Gulick.

Apex leucorraphe Gulick, P. Zool. Soc. London, 1873, p. 79, pl. x, fig. 2.
HAB. Oahu, Kalaikoa.

var. *cinerosa* Pfeiffer.

Achatinella (Bulimella) cinerosa Pfeiffer, P. Zool. Soc. London, 1855, p. 2,
   pl. xxx, fig. 5.

HAB. Oahu.

It is with much diffidence that the above 'lumping' has been attempted. While
not very difficult as a general rule to identify single specimens, I have been unable to
divide the long series of specimens which, partly collected by Mr Perkins and partly by
Mr Hutchison, have passed through my hands. Shape, colour, and form, seem to be
as nothing, and one passes by infinitesimal graduations from one so-called species to
another. True it is that the animals are said to differ in colour, but this alone is not, in
my opinion, sufficient for a specific character; even if it be so, it can, in general, only
avail the field naturalist, and not the Museum student.

The variety *neglecta* Smith is not really so greenish as represented in the original
figure; fig. 23 on the same plate gives a better idea of the real colour.

Férussac appears to have confounded more than one distinct form under his *decora*;
the name has therefore been here used for the species he first referred to.

(8)   *Achatinella dolium* Pfeiffer.

Achatinella (Bulimella) dolium Pfeiffer, P. Zool. Soc. London, 1855, p. 5, pl. xxx,
   fig. 15.

HAB. Molokai (Baldwin).

I fancy this habitat must be wrong and that the species really belongs to Oahu;
the shell is very close to *A. hanleyana* Pf., and may prove to be only a colour variety.

(9)   *Achatinella hanleyana* Pfeiffer.

Achatinella (Bulimella) hanleyana Pfeiffer, P. Zool. Soc. London, 1855, p. 202.

Related to the form of *A. lorata* described as *A. nobilis*, and may prove to be an
extreme variety.

HAB. Oahu.

(10) *Achatinella lorata* Férussac.

Helix (Cochlogena) lorata Férussac, Prodrome, 1822. p. 60.
Achatina lorata Férussac, Deshayes, Hist. Moll. II. p. 193. pl. clv. figs. 9—11.
Achatinella lorata Férussac, Newcomb, Ann. Lyc. New York, IV. p. 310 [animal];
    Semper, Reis. im Philippinen. Landmollusken. pl. xvi. fig. 23 [anatomy].
Achatinella alba Nuttall, Jay, Cat. Shells, Ed. III. 1839. p. 58 [*nomen solum*].
Achatinella pallida Nuttall, Jay, *loc. cit.*; Reeve, Conch. Icon. *Achatinella*, sp. 2.
Achatinella (Bulimella) nobilis Pfeiffer, P. Zool. Soc. London, 1855. p. 202.
Achatinella ventrosa Pfeiffer, *op. cit.* 1855. p. 6. pl. xxx. fig. 20.
Non *A. lorata* Férussac, Reeve, Conch. Icon. *Achatinella*, sp. 6.

A very variable shell, with or without colour bands, and, occasionally, pure white.

Hab. Oahu (various authors); Manoa to Halawa (Baldwin); Nuuanu, Head of Panoa Valley, Mount Tantalus (Perkins).

(11) *Achatinella multilineata* Newcomb.

Achatinella multilineata Newcomb, P. Zool. Soc. London, 1853 [1854], p. 138.
    pl. xxii. fig. 23.
Achatinella (Bulimella) monacha Pfeiffer, *op. cit.* 1855. p. 3. pl. xxx. fig. 9.

Hab. Oahu, Waianae Mountains (Baldwin); Koolau poko (Newcomb). Dr Hartman referred this species, apparently by error, to Maui.

(12) *Achatinella napus* Pfeiffer.

Achatinella (Achatinellastrum) napus Pfeiffer, P. Zool. Soc. London, 1855. p. 5.
    pl. xxx. fig. 19.
Achatinella (Bulimella) concavospira Pfeiffer, *op. cit.* 1859, p. 30.
Apex leucozonus Gulick, *op. cit.* 1873, p. 83, pl. x. fig. 6.

Hab. Oahu.

I regret to be unable to agree with Newcomb that *A. napus* is the same as *A. pulcherrima* Swainson. *A. concavospira* seems to be only an elongate variety; the types of both species are in the British Museum (Natural History).

(13) *Achatinella ovum* Pfeiffer.

Achatinella (Achatinellastrum) ovum Pfeiffer, P. Zool. Soc. London, 1856. p. 334.

Hab. Oahu.

(14)   *Achatinella pulchella* Pfeiffer.

Achatinella (Achatinellastrum) pulchella Pfeiffer, P. Zool. Soc. London, 1855, p. 6, pl. XXX. fig. 2.

A small species, very variable in colour, with a blunt apex, and somewhat depressed in form.

HAB.   Oahu, mountains behind Ewa (Perkins).

(15)   *Achatinella sordida* Newcomb.

Achatinella sordida Newcomb, P. Zool. Soc. London, 1853 [1854], p. 139, pl. XXIII. fig. 27.

Some specimens run very close to *A. decora* Fér.

HAB.   Oahu, Lihue (Newcomb).

(16)   *Achatinella swainsoni* Pfeiffer.

Achatinella (Bulimella) swainsoni Pfeiffer, P. Zool. Soc. London, 1853, p. 4, pl. XXX. fig. 13.

Newcomb suggested that this might be only a form of *A. sordida;* it appears, however, to be distinct, being broader, brown in general coloration, and having a brown, in place of a white lip.   It is a little doubtful, from its form, if it be correctly placed in this group, but the sections are very artificial.

HAB.   Oahu.

(17)   *Achatinella vittata* Reeve.

Achatinella vittata Reeve, Conch. Icon. *Achatinella.* 1850, sp. 9.
Achatinella simulans Reeve, *loc. cit.* sp. 15.
Achatinella (Achatinellastrum) globosa Pfeiffer, P. Zool. Soc. London, 1855, p. 7, pl. XXX. fig. 25.
Apex albofasciatus Smith, *op. cit.* 1873, p. 78, pl. IX. fig. 21.
Apex tumefactus Gulick, *tom. cit.* p. 82, pl. IX. fig. 20.
Helix decora Férussac, Quoy and Gaimard, Voy. Uranie et Phys. 1824. Zool. p. 478 [nec *H. decora* Fér. 1822 = *A. perversa* Swainson].
Achatina decora Férussac, Deshayes, Hist. Moll. II. pt. 2, p. 191. pl. CLV. figs. 5, 7.
Achatinella decora Férussac, Newcomb, Ann. Lyc. New York, VI. p. 307 [animal].
? ? *Achatinella vestita* Mighels, P. Boston Soc. II. 1845, p. 20.

HAB.   Oahu, Waheawa, Halemano, Nuuanu Valley, &c.

var. *cinerea*, n. var.

Banding almost black on the last whorl, ash coloured on the whorl above, the upper whorls tinted with pale brown banding above the suture, replaced by an almost black line at the apex.

HAB. Oahu, Nuuanu (Perkins).

Having examined the types of the first five species mentioned in the above synonymy, I am unable to separate them specifically; with a fair series of specimens the forms shade one into another. The variety is noteworthy for its banding being ashy and almost black, while in the typical form it is red-brown in various patterns. If *A. vestita* be really this species it takes precedence in date: I have never seen a specimen.

subgen. BULIMELLA Pfeiffer.

Bulimella Pfeiffer, Malak. Blätt. 1 (1854), p. 119 (as section of *Achatinella*, first species *A. rosea* Swainson).

(18) *Achatinella (Bulimella) abbreviata* Reeve.

Achatinella abbreviata Reeve, Conch. Icon. *Achatinella*, sp. 19; Newcomb, Ann. Lyc. New York, VI. p. 317 [animal].
Achatinella bacca Reeve, *loc. cit.* sp. 45; Newcomb, *loc. cit.* p. 318 [animal].
Achatinella nivesa Newcomb, P. Zool. Soc. London, 1853 [1854], p. 132, pl. XXII. fig. 6.
Achatinella (Achatinellastrum) clementina Pfeiffer, P. Zool. Soc. London, 1855 [Feb. 1856], p. 205.

HAB. Oahu, Palolo and Konahuanui (Baldwin); Niu (Newcomb); Head of Kawailoa Gulch (Perkins).

The specimens found are of a puzzling form, shewing links between *clementina* and *colorata*.

The animal, as described by Newcomb, seems to vary a good deal in colour.

(19) *Achatinella (Bulimella) ampla* Newcomb.

Achatinella ampla Newcomb, P. Zool. Soc. London, 1853 [1854], p. 137, pl. XXII. fig. 19.

Mr Baldwin considered this a synonym of *A. colorata* Rve.; the only specimen I have seen is the type, which is somewhat injured, and I incline to place it near *A. abbreviata* Rve.

HAB. Oahu, Koolau (Newcomb).

(20) *Achatinella* (*Bulimella*) *bulimoides* Swainson.

Achatinella bulimoides Swainson, Brand's Journ. 1828, p. 85 ; Zool. Illustr. ser. 2, n. pl. CVIII. fig. 1 ; Reeve, Conch. Icon. *Achatinella*, sp. 8 ; Heynemann, Malak. Blätt. XIV. (1867), p. 146, pl. I. fig. 2 [anatomy].
Achatinella obliqua Gulick, Ann. Lyc. New York, VI. p. 245, pl. VIII. fig. 63.
Achatinella oomorpha Gulick, *l. c.*, p. 246, pl. VIII. fig. 64.

A. obliqua was united with this species by Newcomb ; Mr Baldwin, however, gives it as distinct. This latter view may be correct, but the two forms are very closely related.

HAB. Oahu, Kahana (Gulick) ; Kawailoa (Baldwin).

(21) *Achatinella* (*Bulimella*) *byronii* Wood.

Helix byronii Wood, Index Test. Suppl. p. 22, pl. VII. fig. 30.
Achatinella melanostoma Newcomb, P. Zool. Soc. London, 1853 [1854], p. 132, pl. XXII. fig. 7.
Achatinella limbata Gulick, Ann. Lyc. New York, VI. p. 252, pl. VIII. fig. 70.
Achatinella mahogani Gulick, *l. c.*, p. 254, pl. VIII. fig. 72.
Achatinella pulcherrima Swainson, Zool. Ill. pl. CXXIII. fig. 2 ; Gwatkin, P. Ac. Philad. 1895, p. 238 [radula].

HAB. Oahu, Ewa (Newcomb) ; Ahonui, Kalaikoa (Gulick) ; Panoa Valley, Halemano, and ridges between Opaeula and Kawailoa Gulches (Perkins).

var. *recta* Newcomb.

Achatinella recta Newcomb, P. Zool. Soc. London, 1853 [1854], p. 143, pl. XXIII. fig. 45.
Bulimella multicolor Pfeiffer, *op. cit.* 1855, p. 4, pl. XXX. fig. 11 [*pars, non* fig. 11 a].

HAB. Oahu, Waialua (Newcomb) ; Halemano and Nuuanu Valley (Perkins).

var. *nympha* Gulick.

Achatinella nympha Gulick, Ann. Lyc. New York, VI. p. 251, pl. VIII. fig. 69.

HAB. Oahu, Ahonui, Wahiawa, Halemano, Kawailoa, Waimea (Gulick) ; Halemano (Perkins).

The variation is, as usual in the group, very great. A long series collected by Mr Hutchison, added to those of Mr Perkins, has led me to be unable to form any definite break between the various described species which are here placed as varieties. *A. pulcherrima* appears to be a large race in which the colouring has been broken into bands. *A. multicolor* and *A. recta* are, I think, only dwarf varieties. The sinistral shell figured by Pfeiffer (*loc. cit.* pl. xxx. fig. 11 a) as a variety of *A. multicolor* belongs really to *A. oviformis*. *A. nympha* seems a small, elongate, almost colourless variety, with a white lip.

(22) *Achatinella (Bulimella) decipiens* Newcomb.

Achatinella decipiens Newcomb, P. Zool. Soc. London, 1853 [1854], p. 153, pl. xxiv. fig. 68; Ann. Lyc. New York, vi. p. 332 [animal].

Achatinella viridans Pfeiffer, Mal. Blätt. 1854, p. 121 [nec Mighels, fide Newcomb].

Achatinella planospira Pfeiffer, P. Zool. Soc. London, 1855, p. 3, pl. xxx. fig. 8.

Achatinella herbacea Gulick, Ann. Lyc. New York, vi. p. 233, pl. viii. fig. 52.

Achatinella scitula Gulick, *l. c.* p. 241, pl. viii. fig. 61.

Hab. Oahu, Kahana (Newcomb, Baldwin); Koolauloa (Hartman); Waimea, Kawailoa, Hakipu (Gulick).

(23) *Achatinella (Bulimella) faba* Pfeiffer.

Achatinella (Bulimella) faba Pfeiffer, P. Zool. Soc. London, 1859, p. 30.

Hab. Hawaiian Islands.

I cannot trace this species in the Brit. Mus.; it seems not to have been recognized by any recent author.

(24) *Achatinella (Bulimella) glabra* Newcomb.

Achatinella glabra Newcomb, P. Zool. Soc. London, 1853 [1854], p. 139, pl. xxii. fig. 25.

Achatinella fricki Pfeiffer, *op. cit.* 1855, p. 3, pl. xxx. fig. 7.

Achatinella platystyla Gulick, Ann. Lyc. New York, vi. p. 196, pl. vi. fig. 25.

Achatinella wheatleyi Newcomb, MS.

Hab. Oahu, Kawailoa to Hauula (Baldwin); Koolau poko (Newcomb); Kawaiawa (Hartman); Kawailoa (Gulick and Perkins).

Only two dead specimens. I think *A. fricki*, which Newcomb placed with *A. ovata*, really belongs here; fig. 7 a, however, belongs to *A. ovata*. The determination of *A. wheatleyi* is from specimens so named in the Brit. Mus.

(25)　*Achatinella (Bulimella) elegans* Newcomb.

Achatinella elegans Newcomb, P. Zool. Soc. London, 1853 [1854], p. 149,
pl. XXIV. fig. 57.

HAB.　Oahu, Hauula (Newcomb); Hauula and Kaipapau (Baldwin).

(26)　*Achatinella (Bulimella) luteostoma* Baldwin.

Achatinella (Bulimella) luteostoma Baldwin, P. Ac. Philad. 1895, p. 217, pl. x.
figs. 7, 8 [with a note on the animal].

HAB.　Oahu, Palolo to Niu (Baldwin).

(27)　*Achatinella (Bulimella) lymaniana* Baldwin.

Achatinella (Bulimella) lymaniana Baldwin, P. Ac. Philad. 1895, p. 219, pl. x.
figs. 12, 13.

HAB.　Oahu, Waianae mountains (Baldwin).

(28)　*Achatinella (Bulimella) lyonsiana* Baldwin.

Achatinella (Bulimella) lyonsiana Baldwin, P. Ac. Philad. 1895, p. 218, pl. x.
figs. 9—11 [with note on animal]; Suter, *l. c.* p. 239, pl. XI. fig. 52
[radula].

HAB.　Oahu, Konahuanui mountain (Baldwin).

(29)　*Achatinella (Bulimella) ovata* Newcomb.

Achatinella ovata Newcomb, Ann. Lyc. New York, VI. p. 22 [May, 1853]; *T. c.*
p. 324 [animal]; P. Zool. Soc. London, 1853, p. 130, pl. XXII. fig. 2.
Bulimella candida Pfeiffer, P. Zool. Soc. London, 1855, p. 2, pl. XXX. fig. 4.
Achatinella phaeozona Gulick, Ann. Lyc. New York, VI. p. 215, pl. VII. fig. 40.
Achatinella spadicea Gulick, *l. c.* p. 247, pl. VIII. fig. 65.
Achatinella lorata Reeve, Conch. Icon. *Achatinella*, sp. 6 [nec Férussac].

HAB.　Oahu, Kahana, Waianae (Newcomb); Kawailoa (Baldwin); as *A. phaeozona*, Keawaawa, Kailua, Olomana (Gulick); as *A. spadicea*, Kahana (Gulick); Hauula
to Kahana (Baldwin).

Gulick's two species are unknown to me. I follow Newcomb in placing them
here; Mr Baldwin has, however, given them rank as species.

(30)   *Achatinella (Bulimella) oviformis* (Newcomb) Pfeiffer.

Achatinella oviformis Pfeiffer, P. Zool. Soc. London, 1855, p. 208.
Achatinella multicolor Pfeiffer, *t. c.* p. 4, pl. xxx. fig. 11a [nec fig. 11, which
   equals *A. byronii*, var.].

Hab.   Oahu (various authors).

(31)   *Achatinella (Bulimella) rosea* Swainson.

Achatinella bulimoides var. *rosea* Swainson, Brand's Journ. 1828, p. 85.
Achatinella rosea Swainson, Zool. Illustr. ser. 2, pl. cxxiii. fig. 1 ; Reeve,
   Conch. Icon. *Achatinella*, sp. 28 ; Newcomb, Ann. Lyc. New York, vi. p. 309
   [animal].
Bulimella rosea Swainson, Hartman, P. Ac. Philad. 1888, p. 30, pl. i. fig. 4.

A good series, including some varieties approaching *A. ovata*.

Hab.   Oahu, Wahiawa to Kawailoa (Baldwin) ; Waialua (Hartman) ; Halemano
(Perkins).

(32)   *Achatinella (Bulimella) rotunda* Gulick.

Achatinella rotunda Gulick, Ann. Lyc. New York, vi. p. 249, pl. viii. fig. 67.

This form has, with much doubt, been allowed specific rank.   In this I have
followed Mr Baldwin ; Newcomb considered it a variety of *A. ovata*.

Hab.   Oahu, Kaawa and Kahana (Gulick) ; Head of Kawailoa (Perkins).

(33)   *Achatinella (Bulimella) rugosa* Newcomb.

Achatinella rugosa Newcomb, P. Zool. Soc. London, 1853 [1854], p. 138,
   pl. xxii. fig. 22.
Achatinella corrugata Gulick, Ann. Lyc. New York, vi. p. 248, pl. viii. fig. 66.
Achatinella torrida Gulick, *t. c.* p. 250, pl. viii. fig. 68.

I strongly suspect that this will prove to be only a roughened form of *A. byronii*
Wood.

Hab.   Oahu, Ewa (Newcomb) : as *A. corrugata*, Hakipu (Gulick) : Kahana
(Baldwin) : as *A. torrida*, Kahana, Kaawa, Waikane, Waiolu (Gulick).

(34) *Achatinella (Bulimella) sowerbyana* Pfeiffer.

Bulimella sowerbyana Pfeiffer, P. Zool. Soc. London. 1855. p. 4. pl. XXX. fig. 14.

var. *fuscobasis* Smith.

Bulimella fuscobasis Smith, P. Zool. Soc. London. 1873. p. 77. pl. IX. fig. 15.
I think Mr Smith's species is only a colour variety.
HAB. Oahu (type form, authors) ; Mount Kaala (variety, Smith).

(35) *Achatinella (Bulimella) taeniolata* Pfeiffer.

Achatinella taeniolata Pfeiffer, P. Zool. Soc. London, 1846, p. 38 ; Reeve, Conch.
Icon. *Achatinella*, sp. 7.
Achatinella rubiginosa Newcomb, P. Zool. Soc. London, 1854 [1855]. p. 154,
pl. XXIV. fig. 69.
Bulimella macrostoma Pfeiffer, P. Zool. Soc. London. 1855. p. 2, pl. XXX.
fig. 6.
Newcomb was of opinion that *A. macrostoma* was identical with *A. rutila*, but,
after examining the type, I prefer to place it here.
HAB. Oahu, Palolo (Newcomb) ; Niu to Palolo (Baldwin).

(36) *Achatinella (Bulimella) vidua* Pfeiffer.

Bulimella vidua Pfeiffer, P. Zool. Soc. London. 1855. p. 3, pl. XXX. fig. 10.
Newcomb placed this as a synonym of *A. ovata*; Mr Baldwin regarded it as a
distinct species. The columellar plait is very small in the specimens in the Brit. Mus.
HAB. Oahu (Baldwin, &c.).

(37) *Achatinella (Bulimella) viridans* Mighels.

Achatinella viridans Mighels, P. Boston Soc. II. (1845). p. 20 ; Newcomb, P. Zool.
Soc. London, 1854, p. 310 [animal].
Achatinella radiata Pfeiffer, P. Zool. Soc. London. 1845 [1846]. p. 89 ; Reeve,
Conch. Icon. *Achatinella*, sp. 35.

Achatinella subvirens Newcomb, P. Zool. Soc. London, 1853 [1854]. p. 136, pl. xxii. fig. 18.

Achatinella rutila Newcomb, t. c. p. 138, pl. xxii. fig. 21; Op. cit. 1854. p. 310 [animal]; Ann. Lyc. New York, vi. p. 326 [animal].

HAB. Oahu, Niu (Newcomb); Palolo, Niu, Konahuanui (Hartman); Nuuanu to Waialae (Baldwin); Nuuanu, Waialae (Perkins).

subgen. PARTULINA Pfeiffer.

Partulina Pfeiffer, Malak. Blätt. 1. 1854. p. 114.

Pfeiffer had no fixed type for his section, but the species all belong to one group, and I would suggest that his first-named, *A. virgulata* Migh., be treated as the type.

Pease, in his review of the genus in 1869, did not alter the grouping, so far as regards *Partulina*.

Mons. Ancey has recently (P. Malac. Soc. London, iii. 1899, p. 270) proposed to place *P. physa* Newc., and its allies, in a new subgenus *Baldwinia*.

(38) *Achatinella (Partulina) anceyana* Baldwin.

Achatinella (Partulina) anceyana Baldwin, P. Ac Philad. 1895. p. 223, pl. x. fig. 16; Gwatkin, t. c. p. 238 [radula].

HAB. Maui, Makawao (Baldwin).

(39) *Achatinella (Partulina) aptycha* Pfeiffer.

Achatinella (Newcombia) aptycha Pfeiffer, P. Zool. Soc. London, 1855 [March], p. 1, pl. xxx. fig. 1.

HAB. Hawaiian Islands. Probably from Maui.

(40) *Achatinella (Partulina) compta* Pease.

Partulina compta Pease, J. Conchyl. xvii. 1869, p. 175.

Curiously enough some specimens collected on Maui, and sent to me by Mr Baldwin, exactly agree with a specimen from Molokai presented by Pease to the British Museum under this name.

HAB. Molokai (Pease); Kawela (Baldwin)—Maui (Baldwin).

(41) *Achatinella (Partulina) confusa* nom. nov.

Achatinella physa Newcomb, P. Boston Soc. v. 1855, p. 218; Amer. J. Conch. ii.
    1866, p. 214, pl. xiii. fig. 10.
Achatinella (Partulina) physa Newcomb, Baldwin, P. Ac. Philad. 1895, p. 225
    [animal].
Nec *A. physa* Newcomb, 1854, q. v. (p. 316).

An inspection of the figures and descriptions given by Newcomb (P. Zool. Soc.
London, 1853, p. 152, pl. xxiv. fig. 64, and as given above) will, I think, show that
he was confusing two species, under the belief that the first description related only
to a young specimen. It therefore becomes necessary to restrict his name to the
species he first referred to, which unfortunately appears to be the same as *A. hawaiiensis* Baldwin, and to rename the other form, which, it is to be regretted, is the species
universally known as *A. physa*. It may be noted that the habitat originally given by
Newcomb agrees with that of Hamakua given by Mr Baldwin for his *A. hawaiiensis*,
whilst Kohala is a different, but adjoining, district.

HAB.   Hawaii, Kohala (Newcomb).

(42) *Achatinella (Partulina) crassa* Newcomb.

Achatinella crassa Newcomb, P. Zool. Soc. London, 1853 [1854], p. 155, pl. xxiv.
    fig. 71.

HAB.   Lanai (Newcomb); near Koele (Perkins).

(43) *Achatinella (Partulina) dolei* Baldwin.

Achatinella (Partulina) dolei Baldwin, P. Ac. Philad. 1895, p. 221, pl. x. figs. 17,
    18; Suter, t. c. p. 238, pl. xi. fig. 55 [radula].
Belongs to the group of *A. tappaniana* C. B. Ad.; specimens, precisely similar to
some kindly sent me by Mr Baldwin, were identified by Mr Gulick as a variety of his
A. fasciata (= *tappaniana*).

HAB.   Maui, Honomanu (Baldwin).

(44) *Achatinella (Partulina) dubia* Newcomb.

Achatinella dubia Newcomb, Ann. Lyc. New York, vi. p. 23 (May, 1853); P. Zool.
    Soc. London, 1853 [1854], p. 152, pl. xxiv. fig. 65.

HAB.   Oahu, among stones, and Waianae on bushes (Newcomb); Makaha Valley,
Waianae Mts (Perkins).

(45)   *Achatinella (Partulina) dwightii* Newcomb.

Achatinella dwightii Newcomb, Ann. Lyc. New York, VI. p. 145 (Oct. 1855);
   Amer. J. Conch. II. p. 213, pl. XIII. fig. 9; Gwatkin, P. Ac. Philad. 1895,
   p. 238 [radula].

Closely related, apparently, to some of the varieties of *A. redfieldi* Newc.

HAB.   Molokai, Kamalo (Baldwin); Mountains (Perkins).

(46)   *Achatinella (Partulina) fusoidea* Newcomb.

Achatinella fusoidea Newcomb, Ann. Lyc. New York, VI. p. 144 (Oct. 1855); Amer.
   J. Conch. II. 1866, p. 213, pl. XIII. fig. 8.

HAB.   Maui, Haleakala (Newcomb).

(47)   *Achatinella (Partulina) gouldi* Newcomb.

Achatinella gouldi Newcomb, Ann. Lyc. New York, VI. p. 21 (May 1853); P. Zool.
   Soc. London, 1853 [1854], p. 129, pl. XXII. fig. 1.
Achatinella talpina Gulick, Ann. Lyc. New York, VI. p. 212, pl. VII. fig. 38 (Dec.
   1856).
Achatinella myrrhea Gulick, Pfeiffer, Mon. Helic. Viv. IV. p. 517.

HAB.   Maui, on Tutui trees, Wailuku Valley (Newcomb); Wailuku (Gulick).

(48)   *Achatinella (Partulina) grisea* Newcomb.

Achatinella grisea Newcomb, P. Zool. Soc. London, 1853 [1854], p. 153, pl. XXIV.
   fig. 66.

HAB.   Maui, Makawao (Newcomb, &c.).

(49)   *Achatinella (Partulina) hayseldeni* Baldwin.

Partulina hayseldeni Baldwin, Nautilus, X. p. 31, July 1896.
Plate XI. fig. 2.
Belongs to the group of *A. variabilis* Newc.

HAB.   Lanai (Baldwin); Lanaihale, near highest point of Mountains (Perkins).

(50)   *Achatinella (Partulina) horneri* Baldwin.

Achatinella (Partulina) horneri Baldwin, P. Ac. Philad. 1895. p. 224, pl. X. figs.
   20, 21, 22; Gwatkin, t. c. p. 238 [radula].

HAB.   Hawaii, Hamakua (Baldwin).

(51) *Achatinella (Partulina) lignaria* Gulick.

Achatinella lignaria Gulick. Ann. Lyc. New York, vi. p. 209, pl. vii. fig. 35 (Dec. 1856).

Hab. Maui, Wailuku (Gulick).

var. *crocea* Gulick.

Achatinella crocea Gulick, t. c. p. 211, pl. vii. fig. 36 (Dec. 1856).

I think *A. crocea* is only a variety; both were placed by Newcomb as synonyms of his *A. terebra*.

Hab. Maui, Waihee (Gulick).

(52) *Achatinella (Partulina) marmorata* Gould.

Achatinella marmorata Gould, P. Boston Soc. ii. p. 200 (1847); U. S. Explor. Exped. Moll. fig. 94; Newcomb, Ann. Lyc. New York, vi. p. 311 [animal]; Gwatkin, P. Ac. Philad. 1895, p. 238 [radula].

Achatinella adamsi Newcomb, Ann. Lyc. New York, vi. p. 19 (May, 1853); P. Zool. Soc. London, 1853 [1854]. p. 137, pl. xxii. fig. 20 (as *A. adamsii*).

Achatinella induta Gulick, Ann. Lyc. New York, vi. p. 207, pl. vii. fig. 34 (Dec. 1856).

The synonymy of this species is difficult; Newcomb united two other forms described by Gulick from a different district of Maui; Mr Baldwin on the other hand regards them as species. For the present I have left them, with some hesitation, specific rank; they are *A. ustulata* and *A. plumbea*.

Hab. Maui, Haleakala (Gould); Makawao (Newcomb, Baldwin); Wailuku (Gulick).

(53) *Achatinella (Partulina) mighelsiana* Pfeiffer.

Achatinella mighelsiana Pfeiffer, Mon. Hel. Viv. ii. p. 238; Newcomb, Ann. Lyc. New York, vi. p. 319 [animal]; Gwatkin, P. Ac. Philad. 1895, p. 238 [radula]; Reeve, Conch. Icon. *Achatinella*, sp. 40.

The typical form is a whitish shell with a single black band at the periphery; this single band is occasionally split into two smaller ones. Some lovely varieties were collected by Mr Perkins, which may be tabulated as follows:

(α) White and bandless.

(β) Bandless, of a rich orange hue with strigations of a slightly darker shade, tubercle white.

(γ) One-banded, the whitish shell tinted with yellow, ashy, or slaty strigations. A few are white above the band, yellowish below, and show traces of a second band in the umbilical area.

(δ) Two- and even three-banded, ground-colouring white, tinted faintly with ashy strigations, shell not quite so attenuate.

HAB. Molokai, Kalae (Baldwin) ; the Mountains (Perkins).

(54) *Achatinella* (*Partulina*) *morbida* Pfeiffer.

Achatinella (*Bulimella*) *morbida* Pfeiffer, P. Zool. Soc. London, 1859, p. 30.
HAB. ? Oahu.

The only authority I am aware of for the exact habitat is Mr Baldwin, who gives Oahu, but he marks it as one of the species he has not seen.

(55) *Achatinella* (*Partulina*) *mucida* Baldwin.

Achatinella (*Partulina*) *mucida* Baldwin, P. Ac. Philad. 1895, p. 222, pl. x. fig. 23.

A series of about 60 specimens. It is generally of an ashy colour with a dark zone at the periphery ; smaller colour lines are also present in most specimens. The brown stain at the base of the columellar plait is also noteworthy.

HAB. Molokai, Makakupaia (Baldwin) ; Makakupaia, and Mountains of Molokai (Perkins).

(56) *Achatinella* (*Partulina*) *nivea* Baldwin.

Achatinella (*Partulina*) *nivea* Baldwin, P. Ac. Philad. 1895, p. 222, pl. x. fig. 19.
HAB. Maui, Makawao to Huelo (Baldwin).

(57) *Achatinella* (*Partulina*) *perdix* Reeve.

Achatinella perdix Reeve, Conch. Icon. *Achatinella*, sp. 43 (1850) ; Newcomb, Ann. Lyc. New York, vi. p. 317 [animal] ; Gwatkin, P. Ac. Philad. 1895, p. 238 [radula].
Achatinella undosa Gulick, Ann. Lyc. New York, vi. p. 205, pl. vii. fig. 33 (Dec. 1856).

A. undosa was wrongly placed by Clessin (Nom. Helic. Viv. p. 305) as a synonym of *A. radiata* Gould.

HAB. Maui, Lahaina (Baldwin) ; Olinda at 4000 ft. (Perkins).

(58)　*Achatinella* (*Partulina*) *physa* Newcomb.

Achatinella physa Newcomb, P. Zool. Soc. London, 1853 [1854], p. 152, pl. XXIV. fig. 64.

Achatinella (*Partulina*) *hawaiiensis* Baldwin, P. Ac. Philad. 1895, p. 225, pl. X. figs. 24—26; Gwatkin, t. c. p. 238 [radula].

Nec *A. physa* Newc. subsequently.

See for notes on the synonymy under *A. confusa* Sykes.

HAB.　Hawaii, Mauna Kea (Newcomb) ; Hamakua (Baldwin).

(59)　*Achatinella* (*Partulina*) *plumbea* Gulick.

Achatinella plumbea Gulick, Ann. Lyc. New York, VI. p. 213, pl. VII. fig. 39.

HAB.　Maui, Kula (Gulick).

(60)　*Achatinella* (*Partulina*) *porcellana* Newcomb.

Achatinella porcellana Newcomb, P. Zool. Soc. London, 1853 [1854], p. 146, pl. XXIII. fig. 47.

In appearance recalling a dwarf specimen of *A. terebra* Newc. of W. Maui ; only known to me from the type.

HAB.　E. Maui (Newcomb).

(61)　*Achatinella* (*Partulina*) *proxima* Pease.

Helicter proximus Pease, P. Zool. Soc. London, 1862, p. 6.

Partulina proxima Pease, Hartman, P. Ac. Philad. 1888, p. 27, pl. I. figs. 1, 2.

Achatinella proxima Pease, Gwatkin, l. c. 1895, p. 238 [radula].

A fine series. A variety is interesting as showing a link towards *A. theodorei* Baldwin ; it is much more slender and smaller than the typical form, generally lighter in colour, and the colour-markings are much finer in pattern. It was found with the typical form.

HAB.　Molokai, Waikolu (Baldwin) ; Kahanui, and mountains of Molokai (Perkins).

(62) *Achatinella (Partulina) pyramidalis* Gulick.

Achatinella pyramidalis Gulick, Ann. Lyc. New York, VI. p. 204, pl. VII. fig. 32
(Dec. 1856).

Newcomb regarded this as a variety of *A. perdix* Reeve; not having seen specimens which unite them I have left it as a species. Clessin (Nom. Helic. Viv. p.
306) placed it—erroneously—under *A. marmorata* Gould.

Hab. Maui, Lahaina (Gulick); Huelo (Baldwin); Waihee (Perkins).

(63) *Achatinella (Partulina) radiata* Gould.

Achatinella radiata Gould, P. Boston Soc. II. 1845, p. 27.
Bulimus gouldi Pfeiffer, Zeitsch. für Malak. 1846, p. 116.
Partula densilineata Reeve, Conch. Icon. *Partula*, sp. 9.

Hab. ? Maui (Baldwin).

The specimens in the British Museum are labelled "Oahu", but probably this is
erroneous and Maui is the correct habitat.

(64) *Achatinella (Partulina) redfieldi* Newcomb.

Achatinella redfieldi Newcomb, Ann. Lyc. New York, VI. (May 1853), p. 22; t. c.
p. 325 [animal]; P. Zool. Soc. London, 1853 [1854], p. 131, pl. XXII. fig. 5;
Gwatkin, P. Ac. Philad. 1895, p. 238 [radula].

The long series collected by Mr Perkins has given me considerable difficulty.
Newcomb originally gave both Maui and Molokai, Clessin (Nom. Helic. Viv. p. 306)
gave Molokai and Kauai (the latter being obviously wrong), and Mr Baldwin gives
Mapulehu, Molokai. I think Maui was a slip, due to confusion with the very closely
allied *A. splendida*, and that *A. redfieldi* is really a Molokai shell. Next arises the
question of what the typical form may be; Newcomb states that the shell is either plain
or banded on the third whorl *only*, while he gives six as the number of whorls, the shell
figured being banded (as *A. splendida*) on all the whorls. The forms I refer to this
species are:

a. Typical (Plate XI. fig. 15). Varies from nearly white to chestnut, sometimes
being particoloured.

Hab. Makakupaia, Molokai (Perkins).

β. Light to dark fawn colour, banded with brown, the lip being sometimes white. This is the form figured by Newcomb.

HAB. Molokai, towards or above Kamalo (Perkins).

γ. Lip white, shell chestnut, a white band at the periphery and often a smaller one above it, upper whorls finely tessellated. (Plate XI. fig. 16.)

HAB. Molokai, Makakupaia and Kamalo (Perkins).

(65) *Achatinella (Partulina) rufa* Newcomb.

Achatinella rufa Newcomb, Ann. Lyc. New York, VI. p. 21 (May 1853); t. c. p. 324 [animal]; P. Zool. Soc. London, 1853 [1854] p. 130, pl. XXII. fig. 3.

HAB. Molokai, Kalae (Baldwin) : mountains (Perkins).

Dr Hartman gave erroneously, Maui for this shell. The figure is not good, being too elongate and too highly coloured : a pale variety exists.

(66) *Achatinella (Partulina) splendida* Newcomb.

Achatinella splendida Newcomb, Ann. Lyc. New York, VI. p. 20 (May, 1853); P. Zool. Soc. London, 1853 [1854], p. 131, pl. XXII. fig. 4.
Achatinella baileyana Gulick, Ann. Lyc. New York, VI. p. 202, pl. VII. fig. 31 (1856).
Achatinella solida Gulick, Pfeiffer, Mon. Helic. Viv. IV. p. 516.

HAB. Maui, Waihuku (Newcomb, &c.) ; Lahaina and Wailuku (Baldwin).

(67) *Achatinella (Partulina) tappaniana* C. B. Adams.

Achatinella tappaniana C. B. Adams, Contrib. to Conch. p. 126 (1850) [with var. *dubiosa*].
Achatinella eburnea Gulick, Ann. Lyc. New York, VI. p. 199, pl. VI. fig 28; Gwatkin, P. Ac. Philad. 1895, p. 238 [radula].
Achatinella ampulla Gulick, t. c. p. 200, pl. VII. fig. 29.
Achatinella fasciata Gulick, t. c. p. 201, pl. VII. fig. 30.
Achatinella tuba Gulick, Pfeiffer, Mon. Helic. Viv. IV. p. 523.

HAB. Maui, (as *A. tappaniana*) Lahaina (Baldwin); (as *A. eburnea*) Honuaula (Gulick); (as *A. ampulla* and *A. fasciata*) Honukawai (Gulick).

(68)   *Achatinella (Partulina) terebra* Newcomb.

Achatinella terebra Newcomb, P. Zool. Soc. London, 1853 [1854], p. 144, pl.
    xxiii. fig. 40.
Bulimella attenuata Pfeiffer, P. Zool. Soc. London, 1855 [March], p. 4, pl. xxx.
    fig. 12.
Achatinella corusca Gulick, Pfeiffer, Mon. Helic. Viv. iv. p. 525.
Achatinella perforata Gulick, Pfeiffer, pag. cit.

Hab.   Maui; W. Maui (Newcomb); Waihuku (Hartman); Honokowai (Baldwin).

(69)   *Achatinella (Partulina) tessellata* Newcomb.

Achatinella tessellata Newcomb, Ann. Lyc. New York, vi. (May, 1853), p. 19;
    t. c. p. 327 [animal]; P. Zool. Soc. London, 1853 [1854], p. 139, pl. xxiii. fig.
    26; Gwatkin, P. Ac. Philad. 1895, p. 238 [radula].

A very fine series. The forms found at Pelekunu are generally dextral and of
large size; recalling in shape and colouring *A. virgulata*, but as they possess the
mottled colouring of the earlier whorls, so characteristic of the present species, I have
placed them here.

Hab.   Molokai, Kalae to Waikolu (Baldwin); Pelekunu. Makakupaia, Kahanui,
&c. (Perkins).

(70)   *Achatinella (Partulina) ustulata* Gulick.

Achatinella ustulata Gulick, Ann. Lyc. New York, vi. p. 211, pl. vii. fig. 37.
Nec *A. ustulata* Newcomb MS.; fide Pfeiffer, Malak. Blätt. i. p. 136 (= *A.
    colorata* Reeve).

Hab.   Maui, Beautiful Valley (Gulick); Lahaina (Baldwin).

(71)   *Achatinella (Partulina) variabilis* Newc.

Achatinella variabilis Newcomb, P. Zool. Soc. London, 1853 [1854], p. 154, pl.
    xxiv. fig. 70.
Achatinella fulva (Newcomb) Pfeiffer, loc. cit. 1855 [1856], p. 208.
Achatinella lactea Gulick, Ann. Lyc. New York, vi. 1858, p. 198, pl. vi. fig. 27
    [bad].

Hab.   Lanai (Newcomb, &c.); windward side on ridges facing Maui, above
Waiapaa, behind Koele, and Lanaihale (Perkins).

var. *semicarinata* Newc.

Achatinella semicarinata Newcomb, P. Zool. Soc. London. 1853 [1854], p. 156, pl. XXIV. fig. 76.

From an examination of the very fine series collected by Mr Perkins, I think Newcomb was quite right in placing *A. fulva* and *A. lactea* in the synonymy. The former is a straw-coloured form without banding and the latter a white form with a reddish-brown stain in the interior of the aperture. Mr Baldwin remarks that *A. variabilis* is 'invariably dextral,' but sinistral specimens, typical in every other respect, were found by Mr Perkins. In placing *A. semicarinata* as a variety I have been guided by the great difficulty I found in endeavouring to separate this form from *A. fulva*, the type specimens of which shew traces of the carina.

Mr Perkins remarks that 'the broader form with ridge more raised' is 'from higher elevations': it appears to be gradually replaced by the form *fulva* at lower altitudes and this latter shades into *A. variabilis* (typical).

HAB. Lanai (Newcomb, &c.); mountains (Perkins).

(72) *Achatinella (Partulina) virgulata* Mighels.

Partula virgulata Mighels, P. Boston Soc. II. 1845. p. 20.
Achatinella virgulata Mighels, Reeve, Conch. Icon. *Achatinella*, sp. 3; Newcomb, Ann. Lyc. New York, VI. p. 312 [animal]; Gwatkin, P. Ac. Philad. 1895, p. 238 [radula].
Bulimus rohri Pfeiffer, Zeitsch. f. Malak. 1846, p. 115.
Bulimus insignis Mighels, Reeve, Conch. Icon. *Achatinella*, sp. 3.

HAB. Molokai, Kaluaaha to Halawa (Baldwin); Mapulehu and mountains (Perkins).

It is a very variable species and the following, which I take to be a variety, is perhaps worthy of note.

var. α. Either entirely white or slightly tinted with brown on the last whorl; mouth varying from dusky to white; the spiral black line on the upper whorls either present or absent.

HAB. Molokai, Pelekunu (Perkins).

subgen. ACHATINELLASTRUM Pfeiffer.

Achatinellastrum Pfr., Malak. Blätt. I. (1854), p. 133 (first species *A. venulata* Newc.).

(73)   *Achatinella* (*Achatinellastrum*) *augusta* Smith.

Achatinella augusta Smith, P. Zool. Soc. London, 1873, p. 74, pl. IX. fig. 7.
Dr Hartman referred this shell, as *A. angusta*, to *A. fulgens* Newc.
HAB.   Oahu, Waialae, Waialupe, Palolo (Smith).

(74)   *Achatinella* (*Achatinellastrum*) *bella* Reeve.

Achatinella bella Reeve, Conch. Icon. *Achatinella*. sp. 17 ; Newcomb, Ann.
   Lyc. New York, VI. p. 316 [animal]; Gwatkin, P. Ac. Philad. 1895, p. 238
   [radula].
Pease (P. Zool. Soc. 1869, p. 652) united the species, I think erroneously, with
A. polita Newc.
HAB.   Molokai (various authors and Perkins) ; Kalae to Waikolu (Baldwin).

(75)   *Achatinella* (*Achatinellastrum*) *bilineata* Reeve.

Achatinella bilineata Reeve, Conch. Icon. *Achatinella*. sp. 22.
Achatinella johnsoni Newcomb, P. Zool. Soc. London, 1854, p. 147, pl. XXIII. fig. 50.
Achatinella aplustre Newcomb, t. c. p. 147, pl. XXIII. fig. 51.
HAB.   Oahu, Koolau (Newcomb) ; Manoa to Nuuanu (Baldwin).

(76)   *Achatinella* (*Achatinellastrum*) *buddii* Newcomb.

Achatinella buddii Newcomb, P. Zool. Soc. London, 1853 [1854], p. 155, pl. XXIV.
   fig. 73.
Achatinella pexa Gulick, Ann. Lyc. New York, VI. p. 197, pl. VI. fig. 26.
Achatinella plumata Gulick, t. c. p. 217, pl. VII. fig. 41.
Achatinella cæsia Gulick, t. c. p. 234, pl. VIII. fig. 53.
Achatinella fuscozona Smith, P. Zool. Soc. London, 1873, p. 76, pl. IX. fig. 9.
I follow Mr Baldwin in uniting Mr Smith's species, with which I am unacquainted.
Dr Hartman (P. Ac. Philad. 1888) places it (on p. 32) amongst the synonyms of *A. buddii*; possibly this may be a slip as further on (p. 33) he leaves it specific rank, remarking 'this may be a good species, though it approaches very near to *A. fuscolineata*, Smith,' a comparison which appears to me inaccurate.
   HAB.   Oahu, Palolo (Newcomb); Niu, Wailupe, Waialae, Palolo, Kailua, and Waimea (Gulick); Makiki, Palolo (Smith).

(77) *Achatinella (Achatinellastrum) casta* Newcomb.

Achatinella casta Newcomb, P. Zool. Soc. London, 1853, p. 134, pl. xxii. fig. 12.
Achatinella juncea Gulick, Ann. Lyc. New York, vi. p. 230, pl. vii. fig. 49.
Achatinella cognata Gulick, t. c. p. 240, pl. viii. fig. 60.

A. cognata is only known to me from the description: I incline to think Newcomb was right in suppressing it as a species; Mr Baldwin, however, considers it distinct.

Hab.   Oahu, Ewa (Newcomb, Baldwin); Kalaikoa, Wahiawa, Halemano, Haikipuu, and Waikane (Gulick); above Ewa (Perkins).

(78) *Achatinella (Achatinellastrum) cervina* Gulick.

Achatinella cervina Gulick, Ann. Lyc. New York, vi. p. 241, pl. viii. fig. 62.

Newcomb placed it as a variety of *A. ovata*; Mr Baldwin, on the other hand, gives it rank as a species and places it in *Achatinellastrum*. If the specimens in the Brit. Mus. are correctly identified, it is very close to *A. buddii* Newc.

Hab.   Oahu, Kahana (Gulick).

(79) *Achatinella (Achatinellastrum) colorata* Rve.

Achatinella colorata Reeve, Conch. Icon. *Achatinella*, sp. 18; Newcomb, Ann.
     Lyc. New York, vi. p. 316 [animal].
Hab.   Oahu, Ahuimanu (Hartman); Kalihi (Baldwin).

(80) *Achatinella (Achatinellastrum) concolor* Smith.

Achatinella concolor Smith, P. Zool. Soc. London, 1873, p. 75, pl. ix. fig. 1.
Dr Hartman considered it to be a form of *A. colorata* Rve.
Hab.   Oahu, Ewa (Smith).

(81) *Achatinella (Achatinellastrum) cucumis* Gulick.

Achatinella cucumis Gulick, Ann. Lyc. New York, vi. p. 225, pl. vii. fig. 45.
Hab.   Oahu, Kalihi (Gulick); Kalihi to Moanalua (Baldwin); Kaliua (sic) (Hartman).

(82) *Achatinella* (*Achatinellastrum*) *cuneus* Pfeiffer.

Achatinella (*Achatinellastrum*) *cuneus* Pfeiffer, P. Zool. Soc. London, 1855, p. 205.

Newcomb considered this a form of *A. decipiens*; Dr Hartman appears to have been in some confusion, as he placed it (P. Ac. Philad. 1888) at p. 29 under *A. decipiens*, and at p. 30 under *A. viridans*. I have seen a long and characteristic series found on the Island of Oahu by Mr Hutchison.

Hab. Oahu (authors); Halawa (Baldwin); mountains behind Ewa (Perkins).

(83) *Achatinella* (*Achatinellastrum*) *curta* Newc.

Achatinella curta Newcomb, P. Zool. Soc. London, 1853, p. 144, pl. xxiii. fig. 43.
Achatinella undulata Newcomb, P. Boston Soc. v. (1855), p. 219; Amer. J. Conch. ii. (1866), p. 216, pl. xiii. fig. 15; Pfeiffer, P. Zool. Soc. London, 1855, p. 208.
Achatinella dimorpha Gulick, Ann. Lyc. New York, vi. p. 236, pl. viii. fig. 56.
Achatinella albescens Gulick, t. c. p. 237, pl. viii. fig. 57.
Achatinella contracta Gulick, t. c. p. 239, pl. viii. fig. 59.
Achatinella rhodoraphe Smith, P. Zool. Soc. London, 1873, p. 75, pl. ix. fig. 10; Gwatkin, P. Ac. Philad. 1895, p. 238 [radula].
Achatinella pygmaea Smith, P. Zool. Soc. London, 1873, p. 75, pl. ix. fig. 11.

Hab. Oahu, Waialua (Newcomb); various localities (Gulick); Halemano, Waipio, &c. (Smith); between Kawailoa and Waala gulches, generally between Kawailoa and Halemano, Waimea (Perkins).

(84) *Achatinella* (*Achatinellastrum*) *delta* Gulick.

Achatinella delta Gulick, Ann. Lyc. New York, vi. p. 231, pl. viii. fig. 50.

Newcomb considered *A. delta* to be a more banded variety of *A. curta* Newc.; from the material I have seen I incline, with doubt, to leave them distinct.

Hab. Oahu, Kalaikoa, Halemano, &c. (Gulick).

(85) *Achatinella* (*Achatinellastrum*) *diluta* Smith.

Achatinella diluta Smith, P. Zool. Soc. London, 1873, p. 75, pl. ix. fig. 14.

It is near to, but seems distinct from, *A. ligata* Smith, with which Dr Hartman placed it.

Hab. Oahu, probably (Smith).

(86)   *Achatinella (Achatinellastrum) ernestina* Baldwin.

Achatinella (Achatinellastrum) ernestina Baldwin, P. Ac. Philad. 1895, p. 217, pl. x. figs. 5, 6 [animal described].

HAB.   Oahu, Nuuanu Valley (Baldwin).

(87)   *Achatinella (Achatinellastrum) formosa* Gulick.

Achatinella formosa Gulick, Ann. Lyc. New York, VI. p. 235, pl. VIII. fig. 55.

HAB.   Oahu, Waimea (Gulick).

(88)   *Achatinella (Achatinellastrum) fulgens* Newc.

Achatinella fulgens Newcomb, P. Zool. Soc. London, 1853 [1854], p. 131, pl. XXII. fig. 24.

HAB.   Oahu, Niu (Newcomb); Waialua, south-east end (Hartman).

(89)   *Achatinella (Achatinellastrum) germana* Newcomb.

Achatinella germana Newcomb, P. Zool. Soc. London, 1853 [1854], p. 151, pl. XXIV. fig. 61.

HAB.   Maui, Makawao (Newcomb).

(90)   *Achatinella (Achatinellastrum) juddii* Baldwin.

Achatinella (Achatinellastrum) juddii Baldwin, P. Ac. Philad. 1895, p. 216, pl. x. figs. 3, 4.

HAB.   Oahu, Halawa (Baldwin).

(91)   *Achatinella (Achatinellastrum) lehuiensis* Smith.

Achatinella lehuiensis Smith, P. Zool. Soc. London, 1873, p. 76, pl. IX. fig. 4.

I have not seen the species, but it appears from the figure to be near *A. zonata* Gulick; Dr Hartman has suggested that it is a form of *A. multicolor* Pfr. (= *oviformis* Pfr.).

HAB.   Oahu, Lehui (Smith).

(92) *Achatinella (Achatinellastrum) ligata* Smith.

Achatinella ligata Smith, P. Zool. Soc. London, 1873, p. 76, pl. ix. fig. 13.
Achatinella bellula Smith, t. c. p. 77, pl. ix. fig. 8.

I fancy these two forms are only varieties of one species; they approach *A. nympha* Gulick.

Hab. Oahu, Waimolu (Smith); Panoa and Nuuanu (Baldwin); ridges round Nuuanu, Waimea, and beyond head of Panoa Valley (Perkins).

(93) *Achatinella (Achatinellastrum) livida* Swainson.

Achatinella livida Swainson, Zool. Ill. pl. cviii. fig. 2.
Achatinella emmersonii Newcomb, P. Zool. Soc. London, 1853 [1854], p. 156, pl. xxiv. fig. 74.
Achatinella viridans Reeve, Conch. Icon. *Achatinella*, sp. 25 [nec Mighels].
Achatinella reevei C. B. Adams, Contrib. to Conch. 1850. p. 128.
Achatinella consanguinea Smith, P. Zool. Soc. London, 1873, p. 73, pl. ix. fig. 3.
Nec *A. livida* Pfeiffer, P. Zool. Soc. London, 1845, p. 89 [= *A. vulpina* Fér.].

According to Dr Hartman, Mr Smith's species is probably a variety of *A. colorata*; from the specimens I have seen, I think it rather belongs here.

Hab. Oahu, Waialua (Newcomb, Baldwin); Ahuimanu (Smith).

(94) *Achatinella (Achatinellastrum) longispira* Smith.

Achatinella longispira Smith, P. Zool. Soc. London, 1873, p. 73, pl. ix. fig. 5.

Placed by Dr Hartman as a synonym of *A. vulpina*, but the present species is much more slender in form; I should be inclined rather to refer it to the group of *A. olivacea*.

Hab. Oahu, Halawa, Ahuimanu (?) (Smith).

(95) *Achatinella (Achatinellastrum) multizonata* Baldwin.

Achatinella (Achatinellastrum) multizonata Baldwin, P. Ac. Philad. 1895, p. 215, pl. x. figs. 1, 2 [animal described].

The shells collected by Mr Perkins from 'round Nuuanu' are in no sense typical of this species, they appear to be forms shewing links between it and *A. bellula* Smith (= *ligata* Smith); indeed the two may prove to be forms of one variable species.

Hab. Oahu, Nuuanu Valley (Baldwin); ridges round Nuuanu and Waimea (Perkins).

(96)　*Achatinella* (*Achatinellastrum*) *nattii* Baldwin and Hartman.

Achatinella nattii Baldwin and Hartman in Hartman, P. Ac. Philad. 1888, p. 34,
　　pl. 1. fig. 3 [as *nealii* in explanation of plate]; Gwatkin, l. c. 1895, p. 238.
　　[radula].

Hab.　Maui, Makawao to Honomu (Baldwin).

(97)　*Achatinella* (*Achatinellastrum*) *olivacea* Reeve.

Achatinella olivacea Reeve, Conch. Icon. *Achatinella*. sp. 20.
Achatinella prasina Reeve, l. c. sp. 27.

Hab.　Oahu, Manoa to Nuuanu (Baldwin); Nuuanu and Mt. Tantalus (Perkins).

(98)　*Achatinella* (*Achatinellastrum*) *papyracea* Gulick.

Achatinella papyracea Gulick, Ann. Lyc. New York, vi. p. 229, pl. vii. fig. 48.

Hab.　Oahu, Kalaikoa, Ahonui, Wahiawa (Gulick).

(99)　*Achatinella* (*Achatinellastrum*) *polita* Newcomb.

Achatinella polita Newcomb, Ann. Lyc. New York, vi. p. 24 (1853, May); t. c.
　　p. 328 [animal].
Pease considered this to be identical with *A. bella*.

Hab.　Molokai (Newcomb); Kaluaaha to Halawa (Baldwin).

(100)　*Achatinella* (*Achatinellastrum*) *producta* Reeve.

Achatinella producta Reeve, Conch. Icon. *Achatinella*. sp. 13; Newcomb, Ann.
　　Lyc. New York, vi. p. 315 [animal]; Bland and Binney, Ann. Lyc. New
　　York, x. p. 336, pl. xv. figs. 2, 4 [radula and anatomy].
Achatinella venulata Newcomb, P. Zool. Soc. London, 1854. p. 146, pl. xxiii.
　　fig. 48.
Achatinella hybrida Newcomb, t. c. p. 147, pl. xxiii. fig. 52.
Achatinella dunkeri (Cuming MS.) Pfeiffer, op. cit. 1855. p. 208.

Hab.　Oahu, Koolau (Newcomb, Hartman); Manoa to Nuuanu (Baldwin).

(101) *Achatinella (Achatinellastrum) saccata* Pfeiffer.

Achatinella (Achatinellastrum) saccata Pfeiffer, P. Zool. Soc. London, 1859, p. 30.
Unknown to me.
HAB. Hawaiian Isles (Pfeiffer): Oahu (?) (Baldwin).

(102) *Achatinella (Achatinellastrum) solitaria* Newcomb.

Achatinella solitaria Newcomb, P. Zool. Soc. London, 1853 [1854]. p. 130,
  pl. XXIV. fig. 60.
HAB. Oahu, Palolo (Newcomb).

(103) *Achatinella (Achatinellastrum) trilineata* Gulick.

Achatinella trilineata Gulick, Ann. Lyc. New York, VI. p. 226, pl. VII. fig. 46.
HAB. Oahu, Palolo, Waialae, Wailupe, and Niu (Gulick).

(104) *Achatinella (Achatinellastrum) versipellis* Gulick.

Achatinella versipellis Gulick, Ann. Lyc. New York, VI. p. 224, pl. VII. fig. 44.
HAB. Oahu, Kailua (Gulick).

(105) *Achatinella (Achatinellastrum) vulpina* Férussac.

Helix vulpina Ferussac, Voy. Freycinet, Zool. p. 447. pl. LXVIII. figs. 13, 14;
  Souleyet, Voy. Bonite, Zool. II. p. 508, pl. XXIX. figs. 3, 4 [animal].
Achatina vulpina Férussac, Deshayes, Hist. Moll. II. pt. 2, p. 193, pl. CLV. fig. 1.
Achatinella vulpina Reeve, Conch. Icon. *Achatinella*, sp. 29.
Achatinella castanea Reeve, l. c. sp. 24.
Achatinella adusta Reeve, l. c. sp. 30.
Achatinella virens Gulick, Ann. Lyc. New York, VI. p. 254, pl. VIII. fig. 73 (1858).
Achatinella fuscolineata Smith, P. Zool. Soc. London, 1873, p. 75, pl. IX. fig. 2.
Achatinella livida Pfeiffer, P. Zool. Soc. London, 1845, p. 89 [nec Swainson].

   HAB. Oahu, Palolo (Baldwin); Kailua, Palolo, Halawa (Smith); Manoa to
Nuuanu (Baldwin); Nuuanu Valley and Mt. Tantalus (Perkins).

var. *stewarti* Green.

Achatina stewarti Green, Contrib. Macl. Lyc. Philad. i. (1827, July), p. 47, pl. iv.
figs. 1—4.

Achatinella stewarti Green, Reeve, Conch. Icon. *Achatinella*, sp. 26.

Achatinella pulcherrima Reeve, l. c. sp. 23, fig. *a* [nec Swainson].

Achatinella tricolor Smith, P. Zool. Soc. London, 1873, p. 70, pl. ix. fig. 6.

HAB. Oahu (various authors); Heia (Smith); Nuuanu and Mt. Tantalus
(Perkins).

var. *crassidentata* Pfeiffer.

Achatinella (Achatinellastrum) crassidentata Pfeiffer, P. Zool. Soc. London, 1855,
p. 6, pl. xxx. fig. 23.

Achatinella diversa Gulick, Ann. Lyc. New York, vi. p. 220, pl. vii. fig. 42
(1856, Dec.).

Achatinella varia Gulick, t. c. p. 222, pl. vii. fig. 43.

Achatinella analoga Gulick, t. c. p. 227, pl. vii. fig. 47.

HAB. Oahu, Halawa (Baldwin); Halawa, Palolo, Waialae and Wailupe (Gulick);
Waialae and Nuuanu (Perkins).

var. *liliacea* Pfeiffer.

Achatinella (Achatinellastrum) liliacea Pfeiffer, P. Zool. Soc. London, 1859,
p. 31.

HAB. Oahu (Baldwin).

The difficulty of arriving at a satisfactory dividing line between *A. vulpina* and
A. producta is very great. As at present arranged, *A. vulpina* is the brown shell,
var. *stewarti* the greenish coloured form, var. *crassidentata* the parti-coloured, and
var. *liliacea* the bandless variety; all the above being sinistral. *A. producta* on the
other hand is reserved for the larger and, usually, dextral form.

(106) *Achatinella (Achatinellastrum) wailuaensis*, sp. nov.

Testa dextrorsa, subperforata, nitida, turrita, solidula, levissime striata, alba, lineis
castaneis picta, apud peripheriam zona alba, in sutura linea nigro-castanea notata;
anfr. 5—5½, regulariter crescentes, convexi; apertura auriformis; margine columellari
plica fusca mediocri munita, margine dextro acuto, callo parietali tenuissimo. Long.
15·5, alt. 8·4 mill. Plate XI. fig. 19.

A pretty little shell of the group of *A. bella* Reeve, of Molokai. A variety also
occurred (Plate XI. fig. 20) in which the banding is almost obsolete, save in the suture
of the earliest whorls and in one strong dark band below the periphery.

HAB. Maui, Wailua (Perkins).

(107) *Achatinella (Achatinellastrum) zonata* Gulick.

Achatinella zonata Gulick, Ann. Lyc. New York, vi. p. 237, pl. viii. fig. 58.
Achatinella glauca Gulick, l. c. p. 232, pl. viii. fig. 51.

United by Newcomb with *A. trilineata* Gulick; it appears however to have much flatter whorls, and I follow Mr Baldwin, with some little doubt, in restoring it to specific rank. According to Newcomb, *A. glauca* is a synonym of *A. livida* Swain., but specimens in the Brit. Mus. "named from Gulick's type" as a variety, lead me to place it here.

Hab. Oahu, Waimea, Pupukea, Waialei, Kahuku, Hauula, and Kaawa (Gulick); above Ewa (Perkins).

The following appears to be only a manuscript name :

Achatinellastrum olesonii Baldwin, Cat. Shells Hawaiian Islands, 1893, p. 5.

Hab. Oahu, Nuuanu.

PERDICELLA Pease.

Perdicella Pease, P. Zool. Soc. London, 1869, p. 649.

Pease, unfortunately, having named no type, it becomes necessary to select one and I propose to take *A. helena* Newc. The species come from Maui and Molokai.

(1) *Perdicella fulgurans*, sp. nov.

Testa subperforata, dextrorsa, ovato-turrita, nitida, sub lente lineis spiralibus confertim sculpta, albida, strigis fulgurantibus castaneis elegantissime picta, sutura modice impressa, apice obtusulo ; anfr. $5\frac{1}{2}$, plano-convexi, ultimis $\frac{2}{3}$ longitudinis testae aequans ; apertura ovato-pyriformis, intus lilacina ; peristoma margine dextro simplici, columellari subreflexo ; plica columellaris torta, subprominens, mediocris, rapide ascendens. Long. 16 ; lat. 8 ; long. apert. 8·1 ; lat. apert. 4·9 mill. (Plate XI. fig. 5.)

This very pretty shell is akin to *P. zebrina* Pfr., but may be readily separated from it by its greater size, by being much broader in proportion to the length, and by the colour-pattern being finer in design and more zigzag. The protoconch is brown, then becoming paler with a dark shade near the sutural line. It is the *Partulina zebrina* Pfr. of Mr Baldwin's valuable catalogue.

Hab. E. Maui. Makawao to Huelo (Baldwin); Maui (Hutchison).

(2) *Perdicella helena* Newcomb.

Achatinella helena Newcomb, Ann. Lyc. New York, VI. (May, 1853), p. 27 ; P. Zool. Soc. London, 1853 [1854], p. 151, pl. XXIV. fig. 63.

HAB. Molokai, on Ti-tree (Newcomb) ; Kamalo to Kalae (Baldwin) ; Kalae and Makakupaia (Perkins).

(3) *Perdicella mauiensis* (Newcomb) Pfeiffer.

Achatinella mauiensis (sic, err. typ.) (Newcomb) Pfeiffer, P. Zool. Soc. London, 1855 [1856], p. 207 ; Newcomb, Amer. J. Conch. II. p. 217, pl. XIII. fig. 16.

HAB. Maui, Makawao to Huelo (Baldwin).

(4) *Perdicella minuscula* Pfeiffer.

Achatinella (Newcombia) minuscula Pfeiffer, P. Zool. Soc. London, 1858, p. 22.

HAB. Maui, Lahaina (Baldwin).—Molokai Mts. at 4000 feet (Perkins).

Both these habitats can hardly be correct : I suspect the former may be an error of identification.

(5) *Perdicella ornata* Newcomb.

Achatinella ornata Newcomb, P. Zool. Soc. London, 1853 [1854], p. 149, pl. XXIV. fig. 55.

HAB. Maui ; E. Maui (Newcomb); Lahaina (Baldwin).

(6) *Perdicella theodorei* Baldwin.

Achatinella (Partulina) theodorei Baldwin, P. Ac. Philad. 1895, p. 226, pl. X. fig. 27.

HAB. Molokai, Kawela (Baldwin); Makakupaia and the mountains (Perkins).

(7) *Perdicella zebra* Newcomb.

Achatinella zebra Newcomb, Ann. Lyc. New York, VI. p. 142 [Oct. 1853].

Placed by Dr Hartman, in his list, both in *Achatinellastrum* and *Laminella* !

HAB. E. Maui (Newcomb).

(8) *Perdicella zebrina* Pfeiffer.

Newcombia zebrina Pfeiffer, P. Zool. Soc. London. 1855 [1856], p. 202.
HAB. E. Maui (Baldwin as *P. zebra* Newc.).

NEWCOMBIA Pfeiffer.

Newcombia Pfeiffer, Malak. Blätt. 1. 1854, p. 117 : Pease, P. Zool. Soc. London,
1869, p. 649.

Pfeiffer's list of species was very heterogeneous and included shells of diverse
groups, his first species being *A. helena* Newcomb ; fortunately Pease in 1869 properly
confined the group to the shells it is now used for.

Two sections may be formed : I. Spirally lirate ; *N. lirata*, etc. II. Nearly
smooth, usually more elongate ; *N. cumingi*, etc.

(1) *Newcombia canaliculata* Baldwin.

Achatinella (*Newcombia*) *canaliculata* Baldwin, P. Ac. Philad. 1895, p. 226, pl. x.
figs. 28, 29 ; Gwatkin, t. c. p. 238 [radula].
HAB. Molokai, Halawa (Baldwin).

(2) *Newcombia cinnamomea* Pfeiffer.

Achatinella cinnamomea Pfeiffer, Malak. Blätt. IV. 1857, p. 230.
Achatinella (*Newcombia*) *cinnamomea* Pfeiffer, P. Zool. Soc. London. 1858, p. 22.
Newcombia cinnamomea Pfeiffer. Gwatkin, P. Ac. Philad. 1895, p. 238 [radula].
Conchologically this is very close to *N. cumingi* Newc.
HAB. Molokai, Mapulehu (Baldwin) ; Makakupaia and the mountains (Perkins).

(3) *Newcombia cumingi* Newcomb.

Achatinella cumingi Newcomb. Ann. Lyc. New York, VI. 1853. p. 25 ; P. Zool.
Soc. London. 1853 [1854]. pl. XXIV. fig. 59.
HAB. Maui, Haleakala (Newcomb) ; Lahaina and Makawao (Baldwin).

(4) *Newcombia gemma* Pfeiffer.

Achatinella gemma Pfeiffer, Malak. Blätt. IV. 1857. p. 323.
Achatinella (*Newcombia*) *gemma* Pfeiffer, P. Zool. Soc. London. 1858, p. 22.
Akin to *N. lirata* Pfr., but the sculpture is almost obsolete.
HAB. Molokai Mts. (Perkins).

(5) *Newcombia plicata* (Mighels MS.) Pfeiffer.

Achatinella plicata Pfeiffer, Mon. Helic. Viv. II. 1848, p. 235; Newcomb, Ann.
   Lyc. New York, VI. 1858, p. 312 [animal]; Reeve, Conch. Icon. *Achatinella*,
   sp. 44.
Bulimus liratus Pfeiffer, P. Zool. Soc. London, 1851 [1853]. p. 261.

I cannot trace the supposed description by Mighels in P. Boston Soc. as
Bulimus plicatus.

HAB. Molokai, Kalae (Baldwin); Mountains (Perkins).

(6) *Newcombia perkinsi* Sykes.

Newcombia perkinsi Sykes, P. Malac. Soc. London, II. 1896, p. 130.
(Plate XI. fig. 36.)

HAB. Molokai Mts. (Perkins). A fine series of this handsome shell.

(7) *Newcombia pfeifferi* Newcomb.

Achatinella pfeifferi Newcomb, Ann. Lyc. New York, VI. p. 25 (May, 1853);
   P. Zool. Soc. London, 1853 [1854], p. 150, pl. XXIV. fig. 58.
Bulimus newcombianus Pfeiffer, P. Zool. Soc. London, 1851, p. 261 [Dec. 1853].

HAB. Molokai, Kaluaaha (Baldwin).

(8) *Newcombia philippiana* Pfeiffer.

Achatinella philippiana Pfeiffer, Malak. Blätt. IV. 1857, p. 89.

HAB. Molokai, Makakupaia (Baldwin).

(9) *Newcombia sulcata* Pfeiffer.

Achatinella sulcata Pfeiffer, Malak. Blätt. IV. 1857, p. 231.
Achatinella (Newcombia) sulcata Pfeiffer, P. Zool. Soc. London, 1858, p. 22.
Newcombia sulcata Pfeiffer, Gwatkin, P. Ac. Philad. 1895, p. 238 [radula].

HAB. Molokai, Pohakupili (Baldwin).

AMASTRA H. and A. Adams.

Amastra H. and A. Adams, Genera of Recent Mollusca, II. p. 137.

Type: the group of *A. magna* Ad.

This large genus may for convenience be subdivided into groups somewhat in the following manner; perhaps the large first section might be more broken up, though I think no sectional name will prove necessary.

subgen. AMASTRA (s. str.).

(1) *Amastra affinis* Newcomb.

Achatinella affinis Newcomb, P. Zool. Soc. London, 1853 [1854], p. 142, pl. XXIII. fig. 35.

Achatinella (Laminella) goniostoma Pfeiffer, P. Zool. Soc. London, 1855, p. 203.

Amastra rustica Gulick, P. Zool. Soc. London, 1873, p. 84, pl. x. fig. 17.

Dr Hartman has suggested that *A. rustica* may equal *A. variegata* Pfr., an Oahu species.

HAB. E. Maui, Kula (Newcomb, Gulick).

(2) *Amastra albolabris* Newcomb.

Achatinella albolabris Newcomb, P. Zool. Soc. London, 1853 [1854], p. 149. pl. XXIV. fig. 56.

Achatinella nucleola Reeve, Conch. Icon. *Achatinella*, sp. 39 [*non* Gould].

One young specimen I refer to this species. See a note under *A. subrostrata* Pfr.

HAB. Oahu, Waianae Mts. (Newcomb, Perkins); Kapalama and Kalihi (Baldwin).

(3) *Amastra amicta* Smith.

Amastra amicta Smith, P. Zool. Soc. London, 1873, p. 86, pl. x. fig. 20.

Dr Hartman notes that this species "may equal *petricola*"; it appears to me quite distinct.

HAB. Hawaiian Islands (Smith).

(4) *Amastra assimilis* Newcomb.

Achatinella assimilis Newcomb, P. Zool. Soc. London. 1853 [1854], p. 148,
pl. xxiii. fig. 53.
Amastra assimilis Newcomb, Gwatkin, P. Ac. Philad. 1895, p. 238 [radula].
Achatinella deshayesii Morelet, Bull. Soc. Hist. Nat. Moselle, 1857, p. 27
[pars].

It has been suggested that this is a variety of *A. nubilosa* Mighels, but the present
species is a more slender shell with much flatter whorls.

Hab. W. Maui (Newcomb).

(5) *Amastra aurostoma* Baldwin.

Amastra aurostoma Baldwin, Nautilus, x. (July, 1896), p. 31.
Hab. Lanai (Baldwin).

(6) *Amastra badia* Baldwin.

Amastra badia Baldwin, P. Ac. Philad. 1895, p. 230, pl. xi. fig. 40.
Hab. Oahu, Ewa (Baldwin).

(7) *Amastra biplicata* Newcomb.

Achatinella biplicata Newcomb, P. Zool. Soc. London, 1853 [1854], p. 156,
pl. xxiv. fig. 75.
Achatinella deshayesii Morelet, Bull. Soc. Hist. Nat. Moselle, 1857, p. 27 [pars];
Pease, P. Zool. Soc. London, 1869, p. 652.

Morelet's original series, now in the British Museum, consists of three specimens,
one belonging to this species, and two to *A. assimilis* Newc.: his diagnosis however
refers to a form with only one columellar plait. From Mr Perkins' long series, it
appears that the upper plait is variable and sometimes becomes obsolete; in one
specimen, which has received an injury, both plaits are dwarfed so as to show only as a
slight thickening of the columella.

Hab. Lanai (Newcomb); Waiapaa and Koele (Perkins).

(8) *Amastra breviata* Baldwin.

Amastra breviata Baldwin, P. Ac. Philad. 1895, p. 231, pl. xi. figs. 45, 46.
Hab. Oahu, Palolo and Halawa (Baldwin).

(9)   *Amastra citrea* Sykes.

Amastra citrea Sykes, P. Malac. Soc. London, II. (1896), p. 129.
Plate XI. fig. 4.
HAB.   Molokai (Hutchison).

(10)   *Amastra conicospira* Smith.

Amastra conicospira Smith, P. Zool. Soc. London, 1873. p. 86, pl. X. fig. 10.
Dr Hartman places this in the synonymy of *A. assimilis* Newc.; I have never seen the present species, but from the figure it appears distinct.
HAB.   Hawaiian Islands (Smith).

(11)   *Amastra conifera* Smith.

Amastra conifera Smith, P. Zool. Soc. London, 1873. p. 85, pl. X. fig. 11.
HAB.   E. Maui, Kula (Smith).

(12)   *Amastra cornea* Newcomb.

Achatinella cornea Newcomb, P. Zool. Soc. London, 1853 [1854], p. 141, pl. XXIII.
   fig. 32.
HAB.   Oahu, below Kaala (Perkins).
Newcomb appears not to have noted the exact habitat; the type-tablet in the British Museum is, however, labelled 'Oahu.'

(13)   *Amastra crassilabrum* Newcomb.

Achatinella crassilabrum Newcomb, P. Zool. Soc. London, 1853 [1854], p. 141,
   pl. XXIII. fig. 31.
HAB.   Oahu, Waianae (Newcomb, &c.).

(14)   *Amastra cylindrica* Newcomb.

Achatinella cylindrica Newcomb, P. Zool. Soc. London, 1853 [1854]. p. 134,
   pl. XXII. fig. 11 : Ann. Lyc. New York, VI. p. 325 [animal].
HAB.   Oahu, Waianae (Newcomb).

(15)   *Amastra decorticata* Gulick.

Amastra decorticata Gulick, P. Zool. Soc. London, 1873, p. 84, pl. x. fig. 14.

Dr Hartman has united this with *A. ellipsoidea* of Gould, from Maui, but a glance at Gould's figures would have shewn him their distinctness.

HAB.   Oahu, Kawailoa, Halemano, and various localities (Gulick); ridges of Nuuanu (Perkins).

(16)   *Amastra durandi* Ancey.

Amastra durandi Ancey, Naturaliste, 1897, p. 178.
HAB.   Oahu (Ancey).

(17)   *Amastra ellipsoidea* Gould.

Achatinella ellipsoidea Gould, P. Boston Soc. II. (1847), p. 200; U. S. Explor.
Exped. Moll. pl. VII. fig. 96.

A species unknown to me: Newcomb united it with his *A. pupoidea*, but it appears not to be so produced in form.

HAB.   Maui (Gould).

(18)   *Amastra elliptica* Gulick.

Amastra elliptica Gulick, P. Zool. Soc. London, 1873, p. 83, pl. x. fig. 15.

HAB.   Oahu, Waialei, Kahuku, Hauula, Kawailoa (Gulick); Waianae (Hartman).

Two specimens, collected by Mr Perkins on "Waianae Mts. Oahu," may belong to a large, incrassate variety.

(19)   *Amastra extincta* Pfeiffer.

Achatinella (Laminella) extincta Pfeiffer, P. Zool. Soc. London, 1855 [1856],
p. 204.

? ? *Leptachatina hartmani* (Newc. MS.) Hartman, P. Ac. Philad. 1888, p. 54.

I fancy that the new name given by Dr Hartman, on the ground that recent examples had been found, was due to an error of identification.   Specimens submitted to me under the name of *A. extincta* by Mr Baldwin appear to be only a form of *A. similaris* Pease.

HAB.   Oahu, subfossil (Pfeiffer).

(20) *Amastra flavescens* Newc.

Achatinella flavescens Newcomb, P. Zool. Soc. London, 1853 [1854], p. 151, pl.
    XXIV. fig. 62.

HAB. Hawaii (Newcomb).

Dr Hartman gives also " Wanoa, Oahu (Newcomb)," but this must be an error.

(21) *Amastra frosti* Ancey.

Amastra frosti Ancey, Mem. Soc. Zool. France, v. (1892), p. 719 : Sykes, P.
    Malac. Soc. London, III. pl. XIII. fig. 12.

Mons. Ancey has also described (P. Malac. Soc. London, III. p. 269, pl. XII.
fig. 11) a variety *unicolor*.

HAB. Oahu, Waianae (Ancey).

(22) *Amastra grayana* Pfeiffer.

Achatinella (Laminella) grayana Pfeiffer, P. Zool. Soc. London, 1855 [1856, Feb.],
    p. 204.

A single specimen. Dr Hartman has suggested that this is a form of *A. magna*,
but I have not seen linking specimens. It is marked with spiral bands, due probably
to the periostracum.

HAB. Oahu (Clessin. Nomencl. Helic. Viv. 1881) ; ? Oahu (Baldwin).—Lanai,
Lanaihale, a ground shell (Perkins).

(23) *Amastra humilis* Newcomb.

Achatinella humilis Newcomb, Ann. Lyc. New York, VI. (Oct. 1855), p. 143 :
    Amer. J. Conch. II. (1866), p. 211, pl. XIII. fig. 4.

HAB. Molokai, Kalae (Newcomb) ; Makakupaia and the mountains (Perkins).

(24) *Amastra inflata* Pfeiffer.

Achatinella (Laminella) inflata Pfeiffer, P. Zool. Soc. London, 1855 [Feb. 1856]
    p. 203.

HAB. Oahu (Clessin. Nomencl. Helic. Viv.) ; Koolauloa (Baldwin).

(25) *Amastra intermedia* Newcomb.

Achatinella intermedia Newcomb, P. Zool. Soc. London, 1853 [1854], p. 135, pl. XXII. fig. 13.

HAB. Oahu, Waianae (Newcomb); ridges of Nuuanu, and Waianae Mts. below Kaala (Perkins).

A long series.

(26) *Amastra irregularis* Pfeiffer.

Achatinella irregularis Pfeiffer, P. Zool. Soc. London, 1855 [Feb. 1856], p. 205.

HAB. Hawaiian Islands (Pfeiffer).

(27) *Amastra lincolata* Newcomb.

Achatinella lincolata Newcomb, Ann. Lyc. New York, VI. (1853). p. 29: P. Zool. Soc. London, 1853 [1854], p. 140, pl. XXIII. fig. 29.

HAB. Hawaii (Newcomb).

The habitat ' Maui,' originally given, seems to have been a slip.

(28) *Amastra longa* Sykes.

Amastra longa Sykes, P. Malac. Soc. London, II. (Oct. 1896), p. 129.
Plate XI. fig. 35.

HAB. Lanai (Newcomb); windward side, apparently extinct (Perkins).

(29) *Amastra luctuosa* Pfeiffer.

Achatinella (Laminella) luctuosa Pfeiffer, P. Zool. Soc. London, 1855 [Feb. 1856], p. 204.
Laminella luctuosa Pfeiffer, Bland and Binney, Ann. Lyc. New York, X. pp. 335—6 [jaw and radula].

HAB. Oahu, Waialee (Baldwin).

(30)   *Amastra magna* C. B. Adams.

Achatinella magna C. B. Adams, Contrib. to Conch. p. 125. 1850 ; Newcomb,
Ann. Lyc. New York. vi. (1858), p. 319 [animal].
Achatinella baldwinii Newcomb, P. Zool. Soc. London, 1853 [1854], p. 155,
pl. xxiv. fig. 72.

I have seen a good series of this handsome shell.

Hab.   Lanai (Newcomb) ; behind Koele (Perkins).

(31)   *Amastra malleata* Smith.

Amastra malleata Smith, P. Zool. Soc. London, 1873, p. 85, pl. x. fig. 18.

Mr Baldwin has united this with *A. affinis* Newcomb ; I have not seen a specimen,
but, from the figure and description, they appear distinct.

Hab.   E. Maui, Kula (Smith).

(32)   *Amastra mastersi* Newcomb.

Achatinella mastersi Newcomb, P. Zool. Soc. London, 1853 [1854], p. 153, pl.
xxiv. fig. 67 ; Ann. Lyc. New York, vi. p. 332 [animal].
Laminella mastersi Newcomb, Bland and Binney, Ann. Lyc. New York, x. p. 335,
pl. xv, figs. 7, 9--11 [jaw and radula].
Amastra mastersi Newcomb, Gwatkin, P. Ac. Philad. 1895, p. 239 [radula].
Achatinella rubens Pfeiffer, Malak. Blätt. 1854, p. 129 [nec Gould, fide Newcomb].

Hab.   Maui (Newcomb) ; ? Haleakala at 5000 feet (Perkins).

Two, apparently immature, specimens, which I refer here with some hesitation.
Four specimens from " Molokai Mts." appear to me to be very close to this species.

(33)   *Amastra melanosis* Newcomb.

Achatinella melanosis Newcomb, P. Zool. Soc. London, 1853 [1854], p. 144,
pl. xxiii. fig. 41.

Hab.   Hawaii (Newcomb).

(34)   *Amastra modesta* C. B. Adams.

Achatinella modesta C. B. Adams, Contrib. to Conch. 1850, p. 128.
Achatinella pumila Gulick, Clessin, Nom. Helic. Viv. 1881, p. 313.

Hab.   Hawaiian Islands (Adams).—Molokai (Hartman, Baldwin).

(35) *Amastra moesta* Newcomb.

Achatinella moesta Newcomb, P. Zool. Soc. London, 1853 [1854], p. 157, pl. XXIV.
  fig. 77.
Achatinella obscura Newcomb, t. c. p. 157, pl. XXIV. fig. 78.

According to Pease (P. Zool. Soc. London, 1869, p. 651), and he is followed by
Dr Hartman (P. Ac. Philad. 1888, p. 47), these two forms are one species. Probably
this is correct.

HAB. Lanai (Newcomb).

(36) *Amastra mucronata* Newcomb.

Achatinella mucronata Newcomb, Ann. Lyc. New York, VI. (May 1853), p. 28;
  P. Zool. Soc. London, 1853 [1854], p. 146, pl. XXIII. fig. 49.
Achatinella (Laminella) fusiformis Pfeiffer, P. Zool. Soc. London, 1855, p. 5,
  pl. XXX, fig. 18.

HAB. Molokai (Newcomb, Baldwin).

Newcomb, in his later paper, gave the locality of 'Maui,' but probably this is a
slip due to the fact that he was, as he subsequently stated, unable to see the proofs.
Two specimens, collected on Molokai by Mr Perkins, appear to belong to a dwarf race.

(37) *Amastra nana* Baldwin.

Amastra nana Baldwin, P. Ac. Philad. 1895, p. 232, pl. XI. figs. 48, 49 [with
  animal]; Gwatkin, t. c. p. 239 [radula].

HAB. Maui, Makawao at 4000 feet (Baldwin).

(38) *Amastra nigra* Newcomb.

Achatinella nigra Newcomb, P. Boston Soc. V. (Sept. 1855), p. 219; Amer. J.
  Conch. II. (1866), p. 210, pl. XIII. fig. 3.

According to Clessin (Nomencl. Helic. Viv. 1881, p. 311) *A. globosa*, Gulick nec
Pfeiffer, is a synonym.

HAB. E. Maui (Newcomb).

(39) *Amastra nubilosa* Mighels.

Achatinella nubilosa Mighels, P. Boston Soc. II. (1845), p. 20; Reeve, Conch.
  Icon. *Achatinella*, sp. 1; Newcomb, Ann. Lyc. New York. VI. p. 312
  [animal].
Achatinella nubilosa Gould, U.S. Explor. Exped. Moll. pl. VII. fig. 95.
HAB. Molokai (Newcomb); Kalae (Baldwin).
It has been suggested, but I think erroneously, that this species comes, in reality,
from Oahu.

(40) *Amastra nucula* Smith.

Amastra nucula Smith, P. Zool. Soc. London, 1873, p. 85. pl. X. fig. 19.
HAB. Lanai ? (Smith).

(41) *Amastra peasei* Smith.

Amastra peasei Smith, P. Zool. Soc. London, 1873, p. 86. pl. X. fig. 13.
HAB. Hawaiian Islands (Smith).—Oahu ? (Baldwin).

(42) *Amastra pellucida* Baldwin.

Amastra pellucida Baldwin, P. Ac. Philad. 1895, p. 231, pl. XI. figs. 41, 42 [with
  animal].
HAB. Oahu. Waianae Valley (Baldwin).

(43) *Amastra petricola* Newcomb.

Achatinella petricola Newcomb, Ann. Lyc. New York. VI. (Oct. 1855). p. 143;
  Amer. J. Conch. II. (1866), p. 211, pl. XIII. fig. 6.
Achatinella (Laminella) umbilicata Pfeiffer, P. Zool. Soc. London. 1855 [Feb.
  1856]. p. 205.
Amastra umbilicata Pfr., Hartman, P. Ac. Philad. 1888, p. 50. pl. I. fig. 11.
HAB. Molokai (Newcomb); Mapulehu (Baldwin).

(44) *Amastra porphyria* Newcomb.

Achatinella porphyria Newcomb, P. Zool. Soc. London, 1853 [1854], p. 136,
  pl. XXII. fig. 16.
Achatinella (Laminella) grossa Pfeiffer, P. Zool. Soc. London, 1855 [Feb. 1856],
  p. 204.
HAB. Oahu, Waianae (Newcomb).

(45) *Amastra porphyrostoma* Pease.

Amastra porphyrostoma Pease. J. Conchyl. XVII. (1869), p. 172 ; Hartman, P. Ac.
Philad. 1888, p. 48, pl. 1. fig. 6.

HAB. Oahu (Pease).

(46) *Amastra pullata* Baldwin.

Amastra pullata Baldwin, P. Ac. Philad. 1895, p. 228, pl. XI. figs. 31, 32 ;
Gwatkin, t. c. p. 239 [radula].
Amastra umbrosa Baldwin, t. c. p. 229, pl. XI. figs. 36, 37 ; Gwatkin, t. c. p. 239
[radula].

After an examination of the long series collected by Mr Perkins I have been
unable to sever these two species. The animals are said to differ in colour, and they
are said to inhabit different districts ; conchologically they seem to shade into one
another, and the radula appears to be identical. Probably they will prove to be local
races.

HAB. Molokai, Kamalo (Baldwin, as *A. umbrosa*) ; Waikolu (Baldwin, as *A.
pullata*) ; Kamalo and Makakupaia Mts. (Perkins).

(47) *Amastra pupoidea* Newcomb.

Achatinella pupoidea Newcomb, P. Zool. Soc. London, 1853 [1854], p. 144,
pl. XXIII. fig. 42.

The colouring is somewhat exaggerated in the figure. See also a note under *A.
ellipsoidea* Gould.

HAB. E. Maui (Newcomb).

(48) *Amastra pusilla* Newcomb.

Achatinella pusilla Newcomb, Ann. Lyc. New York, VI. (Oct. 1855), p. 144 ;
Amer. J. Conch. II. (1866), p. 211, pl. XIII. fig. 5.
Achatinella pulla (Newcomb) Pfeiffer, P. Zool. Soc. London, 1855 [Feb. 1856],
p. 209.

HAB. Lanai (Newcomb).

(49) *Amastra reticulata* Newcomb.

Achatinella reticulata Newcomb, P. Zool. Soc. London. 1853 [1854], p. 148, pl. xxiv, fig. 54.
Achatinellastrum conspersa Pfeiffer, P. Zool. Soc. London, 1855 [March]. p. 7, pl. xxx. fig. 26.

Hab. Oahu, Waianae (Newcomb).

(50) *Amastra rubens* Gould.

Achatinella rubens Gould, P. Boston Soc. ii. (1845), p. 27 ; Newcomb, Ann. Lyc. New York, vi. p. 314 [animal] ; Reeve, Conch. Icon. *Achatinella*, sp. 42. pars b [fide Newcomb].

Hab. Oahu, W. Mts. (Newcomb) ; Kaala (Baldwin).

(51) *Amastra rubicunda* Baldwin.

Amastra rubicunda Baldwin, P. Ac. Philad. 1895, p. 229, pl. xi. fig. 38 [with animal] ; Suter, t. c. p. 240, pl. xi. figs. 54 [jaw], 56 [radula].

Hab. Oahu, Konahuanui Mt. (Baldwin).

(52) *Amastra rubida* Gulick.

Amastra rubida Gulick, P. Zool. Soc. London, 1873, p. 84, pl. x. fig. 12.

Hab. Oahu, Kahuku (Gulick).

(53) *Amastra sericea* Pfeiffer.

Achatinella (Laminella) sericea Pfeiffer, P. Zool. Soc. London, 1859, p. 31.

Hab. Hawaiian Isles (Pfeiffer).—? Oahu, Waialua (Baldwin). Unknown to me.

(54) *Amastra simularis* Hartman.

Amastra simularis Hartman, P. Ac. Philad. 1888, p. 252, pl. xiii. fig. 7.
Amastra similaris Hartman, Gwatkin, op. cit. 1895, p. 239 [radula].

Hab. Molokai (Hartman, Perkins) ; Mapulehu (Baldwin).

var. *roscotincta* Sykes.

A. simularis Hartman, var. *roscotincta* Sykes. P. Malac. Soc. London. II. p. 130.
Plate XI. fig. 3.
HAB. Molokai mountains (Perkins).

Varieties under the names of *maura* and *semicarnea* have recently been described
by Mons. Ancey (P. Malac. Soc. London, III. p. 270, pl. XIII. figs. 8, 16).

(55) *Amastra solida* Pease.

Amastra solida Pease, J. Conchyl. XVII. (1869), p. 173.
HAB. Oahu (Pease).

(56) *Amastra spirizona* Férussac.

Helix (Cochlogena) spirizona Férussac, Prodrome, no. 433.
Achatina spirizona Fér., Hist. Moll. II. pt. 2, p. 196, pl. CLV. figs. 14, 15.
Achatinella spirizona Fér., Newcomb, Ann. Lyc. New York, VI. p. 307 [animal].
Achatinella acuta Swainson, Quart. J. Sci. Lit. Arts, I. (1828), p. 84; Zool.
  Illustr., ser. 2, pl. XCIX. fig. 3.
Achatinella bætica Mighels, MS.

HAB. Oahu, Waianae Mts. (Baldwin); Waianae Mts. below Kaala on lee side,
and Halemano (Perkins).

var. *nigrolabris* Smith.

Amastra nigrolabris Smith, P. Zool. Soc. London, 1873. p. 85, pl. X. fig. 9.
HAB. Oahu, Wahiawa, Kalaikoa, Waimea (Smith); Halemano (Perkins).

var. *rudis* Pfeiffer.

Achatinella rudis Pfeiffer, P. Zool. Soc. London, 1855, p. 5 (pars).
Laminella albida Pfeiffer, t. c. p. 203.
Newcombia chlorotica Pfeiffer, t. c. p. 203.
HAB. Oahu.

A. nigrolabris Smith, is, in my opinion, only a broader variety, in which the light
band below the suture is wider: in a box of specimens from Halemano forms are found
showing a graduation from it to the typical form. It is possible that *A. rudis* may be
a distinct species, but I fancy it is only a paler colour variety. Mr Perkins notes that
he found the species "mostly on dead branches of trees, covering itself with mucus to
which débris of bark and wood stick, and therefore very well concealed."

(57) *Amastra subrostrata* Pfeiffer.

Achatinella (Laminella) subrostrata Pfeiffer, P. Zool. Soc. London. 1859, p. 31.
HAB. Hawaiian Islands (Pfeiffer).—? Oahu (Baldwin).

Dr Hartman suggests that this is the same as *A. albolabris* Newc. ; it may possibly be an elongate variety, but I doubt it.

(58) *Amastra tenuilabris* Gulick.

Amastra tenuilabris Gulick, P. Zool. Soc. London. 1873, p. 83, pl. x. fig. 16.
HAB. Oahu (Gulick, with some doubt).

Dr Hartman remarks that from "a comparison of types" this is a synonym of *A. flavescens* Newc. I have, equally, examined the types, and this species differs in being stumpier, with a larger mouth, and more shouldered whorls.

(59) *Amastra tenuispira* Baldwin.

Amastra tenuispira Baldwin, P. Ac. Philad. 1895, p. 232, pl. xi. fig. 51.
HAB. Oahu, Kaala Mt. (Baldwin).

(60) *Amastra textilis* Férussac.

Helix (Helicteres) textilis Férussac, Voy. Freycinet, Zool., p. 482.
Achatinella textilis Férussac, Binney, Ann. Lyc. New York. xi. p. 190. pl. xiv. fig. G [radula].
Amastra textilis Férussac, Hartman, P. Ac. Philad. 1888. p. 50. pl. i. fig. 8.
Achatinella microstoma Gould, P. Boston Soc. ii. (1845). p. 28.
Achatinella ventulus Férussac, Reeve, Conch. Icon. *Achatinella*, sp. 31 : Pfeiffer in Conch.-Cab. *Achatinella*, p. 287, pl. lxvii. figs. 12, 13 [nec Férussac].
HAB. Oahu (Baldwin, Hutchison, &c.).

A single specimen from Nuuanu, Oahu, appears to be a varietal form.

(61) *Amastra transversalis* Pfeiffer.

Achatinella (Laminella) transversalis Pfeiffer, P. Zool. Soc. London, 1855 [Feb. 1856]. p. 204.
United by Newcomb with *A. reticulata*, but appears distinct.
HAB. Oahu, Keawaawa (Baldwin).

(62) *Amastra tristis* Férussac.

Helix (Helicteres) tristis Férussac. Voy. Freycinet, Zool. p. 482, pl. LXVIII. figs. 6. 7.

Achatinella tristis Férussac, Reeve, Conch. Icon. *Achatinella*, sp. 37 : Pfeiffer in Conch.-Cab. *Achatinella*, pl. LXVII. figs. 10—11 [not good].

Achatinella fuliginosa Gould, P. Boston Soc. II. (1845), p. 28.

HAB. Oahu, Palolo (Hartman) ; Nuuanu to Palolo (Baldwin).

(63) *Amastra turritella* Férussac.

Helix (Cochlogena) turritella Férussac, Prodrome, No. 434 : Voy. Freycinet, Zool. p. 481.

Achatina turritella Fér., Hist. Moll. II. pt. 2, p. 196, pl. CLV. fig. 13 ; Souleyet, Voy. Bonite. Zool. II. p. 509, pl. XXIX. figs. 7—8 [animal].

Achatinella turritella Fér., Newcomb, Ann. Lyc. New York, VI. p. 307 [animal].

? *Helix (Cochlogena) luteola* Férussac, Voy. Freycinet, Zool. p. 480.

? *Achatina luteola* Fér., Hist. Moll. II. pt. 2, p. 196, pl. CLV. fig. 12.

? *Laminella luteola* Fér. (sic), Hartman, P. Ac. Philad. 1888, p. 42.

Achatina oahuensis Green, Contrib. Macl. Lyc. Phil. I. (1827), p. 49, pl. IV. fig. 5.

Achatinella inornata Mighels, P. Boston Soc. II. (1845). p. 21.

Newcomb states that he was unable to trace the type of *A. luteola*, and apparently it is lost : Pease (P. Zool. Soc. London, 1869, p. 652) united it with *A. turritella*, and probably this will prove to be correct.

HAB. Oahu (authors) ; Kalihi to Palolo (Baldwin) ; ridges of Nuuanu Valley (Perkins).

(64) *Amastra undata* Baldwin.

Amastra undata Baldwin, P. Ac. Philad. 1895, p. 230, pl. XI. fig. 39.

HAB. Oahu, Nuuanu (Baldwin).

(65) *Amastra uniplicata* Hartman.

Amastra uniplicata Hartman, P. Ac. Philad. 1888, p. 50, pl. I. fig. 7.

HAB. Molokai (Hartman).

(66) *Amastra variegata* Pfeiffer.

Achatinella variegata Pfeiffer, Zeitschr. für Malak. 1849, p. 90; Conch.-Cab.
  Achatinella, p. 282, pl. LXVII. figs. 14, 15.
Amastra rubens, Reeve, pars, Conch. Icon. *Achatinella*, fig. 42 a [fide Newcomb].
Achatinella decepta C. B. Adams, Contrib. to Conch. 1850. p. 127.

HAB. Oahu, head of Boothes Valley (Hartman); Waianae (Baldwin).

(67) *Amastra ventulus* Férussac.

Helix (Helicteres) ventulus Férussac, Voy. Freycinet, Zool. p. 481.
Achatinella ventulus Férussac, Newcomb, Ann. Lyc. New York, VI. p. 306
  [animal].
Achatinella melampoides Pfeiffer, P. Zool. Soc. London, 1851 [Dec. 1853]. p. 262;
  Pfeiffer in Conch.-Cab. *Achatinella*, p. 288, pl. LXVII. figs. 8, 9.
Achatinella (Amastra) manoaensis Newc., Clessin, Nom. Helic. Viv. 1881, p. 311.
May prove to be a *Leptachatina*.

HAB. Oahu, Nuuanu to Palolo (Baldwin); Panoa Valley and ridges of Nuuanu
(Perkins).

(68) *Amastra violacea* Newcomb.

Achatinella violacea Newcomb. Ann. Lyc. New York, VI. (1853, May), p. 18;
  P. Zool. Soc. London. 1853 [1854]. p. 135, pl. XXII. fig. 14.
Achatinella gigantea Newcomb. P. Zool. Soc. London, 1853 [1854]. p. 136, pl.
  XXII. fig. 17.

These two species have been united by Pease and Dr Hartman, the latter
remarking "The only example of *gigantea* ever found is in the British Museum. It
probably equals a large example of *A. violacea*, Newc." Probably the locality of Maui,
given by Newcomb, was an error, as his specimen appears to be only an elongate
form of the Molokai shell. This varies very much in size and shape, as may be seen
from the following:

Alt. 34; diam. 16; alt. ap. 15 ; lat. ap. 9·5 mill.
„ 31:   12; „   „ 11·5;   „ 7 „
„ 30:   15; „   „ 9   „   „ 9 „

Mr Baldwin has left *A. gigantea* in his list as a Maui shell, but the fact that this
diligent collector has marked it as a shell unknown to him, lends confirmation to the
view that it does not really come from that island.

HAB. Molokai, Haleakala (Newcomb); Mapulehu to Halawa (Baldwin); Halawa
and Pelekunu (Perkins).—? Maui as *A. gigantea*, Haleakala (Newcomb).

subgenus LAMINELLA Pfeiffer.

Laminella Pfr., Malak. Blätt. I. 1854. p. 126.

Pfeiffer's original group was very heterogeneous, as was that of Pease under this name (P. Zool. Soc. London, 1869. p. 648); the latter author also proposing *Perdicella* for a portion of the group. I would propose to select *A. gravida* Fér., the old and well-known species, as the type.

(69)   *Amastra* (*Laminella*) *alexandri* Newcomb.

Achatinella alexandri Newcomb, P. Calif. Ac. III. (1865). p. 182 ; Amer. J. Conch., II. (1866). p. 216, pl. XIII. fig. 14.

HAB. West Maui, at 7500 feet (Newcomb) ; top of West Maui (Baldwin).

(70)   *Amastra* (*Laminella*) *citrina* (Mighels MS.) Pfeiffer.

Achatinella citrina Mighels. Pfeiffer, Mon. Helic. Viv. II. (1848), p. 234 ; Reeve. Conch. Icon. *Achatinella*, sp. 33 ; Newcomb, Ann. Lyc. New York, VI. p. 312 [animal].

Pease united (P. Zool. Soc. London, 1869, p. 652) this species with *A. venusta*. Conchologically, they differ in the periostracum, shape of whorls, &c., while, from the descriptions given by Newcomb, the animals are distinct in colouration.

HAB. Molokai, Kalae to Waikolu (Baldwin) ; Molokai (Perkins).

(71)   *Amastra* (*Laminella*) *concinna* Newcomb.

Achatinella concinna Newcomb, P. Zool. Soc. London, 1853 [1854]. p. 157, pl. XXIV. fig. 79.

Newcomb's type was a bandless dextral shell. In the very fine series collected, both dextral and sinistral forms occur ; black bands are either absent or present, and, in the latter event, vary from one to even four in number.

HAB. Lanai (Newcomb, &c.) ; Koele side of highest point, side of highest point furthest from Koele, near Koele at 3000 feet (Perkins).

(72)   *Amastra* (*Laminella*) *depicta* Baldwin.

Laminella depicta Baldwin, P. Ac. Philad. 1895, p. 228, pl. XI. figs. 33—5 [animal described].

A very fine series, shewing a range of colour from pale yellow to rich orange, tinged with crimson ; it is sometimes dextral, but sinistral forms predominate.

HAB. Molokai, Kamalo (Baldwin); mountains, and above Pelekunu (Perkins).

(73) *Amastra (Laminella) elongata* Newcomb.

Achatinella elongata Newcomb, Ann. Lyc. New York, VI. (May, 1853), p. 26.
Achatinella acuta Newcomb, P. Zool. Soc. London, 1853 [1854], p. 142 [nec
Swainson].

The figure given in P. Zool. Soc. London (l. c.) under this name does not (fide
Newcomb) represent the present species, but *A. soror.* Dr Hartman gives ' Makawao,
Maui' as the habitat, but this must be an error; further he unites the species,
erroneously in my opinion, with *A. hutchinsonii* Pease.

HAB.   Oahu, Lehue (Newcomb): Waianae Mts (Baldwin).

(74) *Amastra (Laminella) erecta* Pease.

Laminella erecta Pease, J. Conchyl., XVII. (1869), p. 174.
Close to *A. micans* Pfeiffer.
HAB.   Maui (Pease).

(75) *Amastra (Laminella) farcimen* Pfeiffer.

Achatinella (Laminella) farcimen Pfeiffer, P. Zool. Soc. London, 1856, p. 334.
nec *Amastra farcimen*, Pfeiffer, Hartman, P. Ac. Philad. 1888, p. 46, pl. I. fig. 5.

Dr Hartman states that his figure is "typical"; this is obviously incorrect as the
type is a sinistral specimen, of considerable size, while the figure represents a smaller,
dextral, shell of another group.   What species his shell may belong to, I am unable to
determine, but it appears to possess no columellar plait.

HAB.   Maui (Newcomb, fide Pfeiffer).

(76) *Amastra (Laminella) fraterna* Sykes.

Amastra fraterna. Sykes, P. Malac. Soc. London, II. (Oct. 1896), p. 129.
Plate XI. fig. 23.
HAB.   Lanai, mountains behind Koele (Perkins).

(77) *Amastra (Laminella) gravida* Férussac.

Helix gravida Férussac, Voy. Freycinet, Zool. p. 478. pl. LXVIII. figs. 4. 5.
Achatina gravida Férussac. Deshayes, Hist. Moll. II. p. 192, pl. CLV. figs. 3. 4.
Achatinella gravida Fér., Newcomb, Ann. Lyc. New York, VI. p. 307 [animal].

Achatinella suffusa Reeve, Conch. Icon. *Achatinella*, sp. 11.

Achatinella dimondi C. B. Adams, Contrib. to Conch. 1850, p. 126 (with var. *lata*).

The specimen described by Reeve does not now appear to exist in the Brit. Mus. collection.

HAB. Oahu, Kalihi to Niu (Baldwin); Nuuanu (Perkins).

(78) *Amastra (Laminella) helvina* Baldwin.

Achatinella (Laminella) helvina Baldwin, P. Ac. Philad. 1895, p. 227, pl. XI. fig. 30 [shell, animal, and anatomy]; Gwatkin, t. c. p. 239 [radula].

Some specimens, given to Mr Perkins by Mr O. Meyer, are broader and have more periostracum, forming a link towards *A. picta*.

HAB. Molokai, Ohia Valley, near Kaluaaha (Baldwin); Molokai (Perkins).

(79) *Amastra (Laminella) hutchinsonii* Pease.

Helicter hutchinsonii Pease, P. Zool. Soc. London, 1862, p. 7.

Amastra hutchinsonii Pease, Hartman, P. Ac. Philad. 1888, p. 45, pl. I. fig. 9.

Dr Hartman suggests, I think erroneously, that this is a synonym of *A. elongata* Newc.

HAB. Maui (Pease).

(80) *Amastra (Laminella) micans* Pfeiffer.

Achatinella (Laminella) micans Pfeiffer, P. Zool. Soc. London, 1859, p. 31.

Amastra micans Pfeiffer, Hartman, P. Ac. Philad. 1888, pl. I. fig. 10.

Dr Hartman's figure is not very good.

HAB. Oahu (Baldwin, Hutchison).

(81) *Amastra (Laminella) picta* Mighels.

Achatinella picta Mighels, P. Boston Soc. II. (1845). p. 21; Newcomb, Ann. Lyc. New York, VI. p. 311 [animal]; Reeve, Conch. Icon. *Achatinella*, sp. 36; Pfeiffer, Conch.-Cab. *Achatinella*, p. 284, pl. LXVIII. figs. 28, 29 [not very good]; Bland and Binney, Ann. Lyc. New York, X. p. 335, pl. XV. fig. 6 [jaw].

Achatinella picta Pfeiffer, P. Zool. Soc. London, 1845 [1846], p. 90.

HAB. Maui, Lahaina and Makawao (Baldwin); Haleakala, at 4000 feet, and Iao Valley (young shells) (Perkins).

var. *bulbosa* Gulick.

Achatinella bulbosa Gulick, Ann. Lyc. New York. vi. (1858), p. 253. pl. viii. fig. 71.

Newcomb placed *A. bulbosa* as a synonym; it appears to me to be of varietal rank, and to differ in being larger and in the whorls being more flattened. I fancy the species will prove to be variable, as Mr Perkins' shells are more slender than the series in the Museum. Specimens sent by Mr Hutchison as from ' Maui ' are still more slender, and may possibly prove to be distinct. Mighels gave, by error probably, ' Oahu.'

Hab. E. Maui, Honuaula and Kula (Gulick).

(82) *Amastra (Laminella) remyi* Newcomb.

Achatinella remyi Newcomb. Ann. Lyc. New York. vi. (Oct. 1855). p. 146: Amer. J. Conch. ii. (1866). p. 215. pl. xiii. fig. 13.

Hab. Lanai (Newcomb).

Only known to me from the original series in the Brit. Mus.: Pfeiffer (P. Zool. Soc. London. 1855. p. 207) gave Hawaii as the habitat, but probably this was an error.

(83) *Amastra (Laminella) sanguinea* Newcomb.

Achatinella sanguinea Newcomb, P. Zool. Soc. London. 1853 [1854]. p. 135. pl. xxii. fig. 15: Ann. Lyc. New York. vi. p. 326 [animal].
Laminella ferussaci Pfeiffer. P. Zool. Soc. London. 1855 [1856]. p. 203.

Hab. Oahu. Lehui (Newcomb): Waianae and Halemano (Baldwin): Halemano, Kawailoa, and Makaha Valley (dead) (Perkins).

(84) *Amastra (Laminella) soror* Newcomb.

Achatinella soror Newcomb. P. Zool. Soc. London. 1853 [1854]. p. 143. pl. xxiii. fig. 38 [also fig. 36. sub nom. *A. acuta*].

Hab. Maui (Newcomb).

The additional locality of Lanai given. subsequently. by Newcomb. really. I think. refers to my *A. fraterna*.

(85)　*Amastra (Laminella) straminea* Reeve.

Achatinella straminea Reeve, Conch. Icon. *Achatinella*. sp. 38; Newcomb. Ann.
　Lyc. New York, VI. p. 318 [animal].

HAB.　Oahu, Panoa to Palolo (Baldwin); Nuuanu (Perkins).

(86)　*Amastra (Laminella) tetrao* Newcomb.

Achatinella tetrao Newcomb, P. Boston Soc. V. (1855), p. 219; Ann. Lyc. New
　York, VI. p. 334 [animal]; Amer. J. Conch. II. (1866), p. 214, pl. XIII.
　figs. 11, 12.

From the fine series collected it appears that the ground-colouring, under
the zigzag periostracum, varies considerably. Shades of crimson or rich orange predominate, but occasionally the colour is confined to a band below the suture, the
rest of the shell being whitish.

HAB.　Lanai (Newcomb); mountains and behind Koele (Perkins).

(87)　*Amastra (Laminella) venusta* Mighels.

Achatinella venusta Mighels, P. Boston Soc. II. (1845), p. 21; Newcomb, Ann. Lyc.
　New York, VI. p. 311 [animal]; Reeve, Conch. Icon. *Achatinella*. sp. 32;
　Binney, Ann. Lyc. New York, XI. p. 191, pl. XIV. fig. D.

HAB.　Molokai, Mapulehu (Baldwin); mountains (Perkins).

Mighels gave, but erroneously, 'Oahu' as the locality.

(88)　*Amastra (Laminella) villosa* Sykes.

Amastra villosa Sykes, P. Malac. Soc. London, II. (1896), p. 129.

Plate XI. fig. 24.

The specimen here figured is not the one whose measurements were given in the
original diagnosis, but a slightly smaller shell whose periostracum is better preserved.

HAB.　Molokai (Perkins).

subgenus AMASTRELLA, n. subgen.

This name is proposed for a group of rotund, generally incrassated, small forms,
which have been usually placed in *Amastra*. I take as the type *A. rugulosa* Pease.
They are nearly all natives of Kauai, but a few species are found on other islands.

(89)   *Amastra (Amastrella) anthonii* Newcomb.

Achatinella anthonii Newcomb, P. Calif. Ac. II. (1861), p. 93; Amer. J. Conch. II.
(1866), p. 210, pl. XIII. fig. 2.
HAB.   Kauai (Newcomb).

(90)   *Amastra (Amastrella) antiqua* Baldwin.

Amastra antiqua Baldwin, P. Ac. Philad. 1895, p. 233, pl. XI. fig. 47.
HAB.   Oahu, Ewa (Baldwin, as fossil).

(91)   *Amastra (Amastrella) carinata* Gulick.

Amastra carinata Gulick, P. Zool. Soc. London, 1873, p. 83.
Achatinella obesa var. *agglutinans* Newcomb, P. Zool. Soc. London, 1853 [1854],
p. 143, pl. XXIII. fig. 39 a.
This appears to be specifically distinct from *A. obesa* Newc.
HAB.   W. Maui, Wailuku (Gulick).

(92)   *Amastra (Amastrella) cyclostoma* Baldwin.

Amastra cyclostoma Baldwin, P. Ac. Philad. 1895, p. 234, pl. XI. fig. 53 [animal
and shell].
HAB.   Kauai, Makaweli (Baldwin).

(93)   *Amastra (Amastrella) nucleola* Gould.

Achatinella nucleola Gould, P. Boston Soc. II. (1845), p. 28.
Achatinella brevis Pfeiffer, P. Zool. Soc. London, 1845 [1846], p. 90.
HAB.   Kauai (Newcomb); Hanalei (Baldwin).—? Oahu, Manoa Valley (Clessin,
Nomenc. Helic. Viv.).

I feel doubtful as to this last locality; the *A. nucleola* Gould, of Reeve, is
A. albolabris Newc. (cf. p. 333).

46—2

(94) *Amastra* (*Amastrella*) *obesa* Newcomb.

Achatinella obesa Newcomb, Ann. Lyc. New York, VI. (May, 1853). p. 24; t. c.
 p. 320 [animal]; P. Zool. Soc. London, 1853 [1854]. p. 143. pl. XXIII. fig. 39;
 Binney, Ann. Lyc. New York, XI. p. 191, pl. XIV. fig. H [radula and jaw].

HAB. Maui, Makawao and Kula (Baldwin); Haleakala (Newcomb).

(95) *Amastra* (*Amastrella*) *rugulosa* Pease.

Amastra rugulosa Pease. P. Zool. Soc. London, 1869, p. 649 (nom. sol.);
 J. Conchyl. XVIII. (1870), p. 95; Crosse, l. c. XXIV. (1876), p. 99, pl. I. fig. 4.

var. *similaris* Pease.

Amastra similaris Pease, P. Zool. Soc. London, 1869, p. 649 (nom. sol.).
Amastra rugulosa var. *similaris* Pease, J. Conchyl. XVIII. (1870), p. 96.

Mr Perkins' specimens are small but otherwise agree with some presented by Pease
to the British Museum. I have seen specimens collected by Mr Hutchison as from
Oahu, but think there must be an error as to the locality.

HAB. Kauai (Pease, type and var.); Kapaa (Baldwin); Lihue (Perkins).—E.
Maui, Kula (Hartman) [? an error].

(96) *Amastra* (*Amastrella*) *sphaerica* Pease.

Amastra sphaerica Pease. P. Zool. Soc. London, 1869, p. 649 (nom. sol.);
 J. Conchyl. XVIII. (1870), p. 94; Crosse, l. c. XXIV. (1876), p. 98, pl. I.
 figs. 5, 5 *a*.

HAB. Kauai (Pease).

The habitat is given as "? Lanai" by both Mr Baldwin and Dr Hartman, but
I know not on what authority.

(97) *Amastra* (*Amastrella*) *vetusta* Baldwin.

Amastra vetusta Baldwin, Cat. Shells Hawaiian Islands, 1893. p. 10 (nom. sol.);
 P. Ac. Philad. 1895, p. 233, pl. XI. fig. 50.

HAB. Oahu, near the base of Punchbowl Hill, Honolulu, fossil (Baldwin).

subgen. KAUAIA, nom. nov.

Carinella Pfr. (1875) nec Sowerby (1839).

The type of Pfeiffer's group is *A. kauaiensis* Newc. : the subgeneric name was used first by Sowerby for a different group of Molluses. Whether *A. alata* and *A. heliciformis* really belong here I am not clear.

(98) *Amastra (Kauaia) alata* Pfeiffer.

Helix alata Pfeiffer, P. Zool. Soc. London. 1856, p. 33.

I have elsewhere (P. Malac. Soc. London. II. p. 127) pointed out that all authors have overlooked the fact that this shell has a columellar plait. It is, in my opinion, not a Helicoid at all, but belongs to an aberrant group of *Amastra*. The columellar plait does not ascend rapidly into the shell, but stands almost horizontally, and has no final 'knob.' The single specimen found by Mr Perkins measures diam. max. 8; alt. 4; alt. apert. 3 mill.

HAB. Lanai (Pfeiffer) ; Mts. behind Koele (Perkins).

(99) *Amastra (Kauaia) heliciformis* Ancey.

Amastra heliciformis Ancey, Bull. Soc. Malac. France, VII. (1890), p. 340.

HAB. Oahu, Waianae (Ancey).

(100) *Amastra (Kauaia) kauaiensis* Newcomb.

Achatinella kauaiensis Newcomb, Ann. Lyc. New York, VII. (April, 1860), p. 145; Amer. J. Conch. II. (1866), p. 209, pl. XIII. fig. 1.
Achatinella (Carinella) kauaiensis Newc., Pfeiffer, Novit. Conch. IV. p. 115, pl. CXXVI. figs. 8—11.

A good series, principally however dead shells, of this almost extinct species. Mr Perkins notes that one specimen was found "with embryonic shells in mouth."

HAB. Kauai (authors) ; Halemanu (Baldwin) ; Makaweli at 2000 ft. and Halemanu at 4000 feet (Perkins).

(101) *Amastra (Kauaia) knudseni* Baldwin.

Amastra knudseni Baldwin, P. Ac. Philad. 1895, p. 234, pl. xi. figs. 43, 44.

Hab. Kauai, Halemanu (Baldwin, Perkins). A single specimen of this very fine species.

The following appear to be only MS. names: *Amastra ferruginea* Baldwin, Cat. Shells Hawaiian Islands, 1893, p. 9 (nom. sol.). Hab. Oahu, Ewa and Waianae (Baldwin).—*Amastra testudinea* Baldwin, t. c. p. 10. Hab. Oahu, Ewa (Baldwin).

LEPTACHATINA Gould.

Leptachatina Gould, P. Boston Soc. II. p. 201: type *Achatinella acuminata* Gould.

It is frequently difficult to draw the line between this group and *Amastra*, and perhaps such species as *A. melampoides* Pfr. (= *A. ventulus* Fér.) may eventually be transferred to *Leptachatina*.

Pfeiffer proposed *Labiella* (Malak. Blätt. 1 1854. p. 142) for the group with an incrassated lip, such as *A. labiata* Newc., and perhaps it may, conchologically, form a convenient section.

The species are principally from Oahu, but an elongate and generally striate group characterizes the older Islands, such as Kauai.

(1) *Leptachatina accincta* Mighels.

Achatina acciucta (err. typ.) Mighels, P. Boston Soc. II. (1845), p. 20: Reeve, Conch. Icon. *Achatina*, sp. 101.

nec *Achatinella accincta* Gould, U.S. Explor. Exped. Mollusca, pl. VII. fig. 97.

Achatinella granifera Gulick, Ann. Lyc. New York, VI. (1856), p. 185, pl. VI. fig. 13; Sykes, P. Malac. Soc. London, III. pl. XIV. fig. 5.

Achatinella (Leptachatina) margarita Pfeiffer, P. Zool. Soc. London, 1855, p. 206.

Gulick admitted (P. Zool. Soc. London, 1873, p. 91) the identity of his species with Pfeiffer's. If Mighels' dimensions and description are accurate, I think the above identification will prove correct. The shell figured by Gould does not appear to be Mighels' species. See also a note under *L. grana* Newc.

Hab. Oahu (Mighels, Pfeiffer); Keawaawa (Gulick).

(2) *Leptachatina acuminata* Gould.

Achatinella acuminata Gould, P. Boston Soc. II. (1847), p. 200; U. S. Explor.
   Exped. Mollusca, pl. VII. fig. 100.
Plate XII. figs. 13, 13 a.

The type of the genus; the radula is figured from a dissection by Lt.-Col H. H.
Godwin-Austen, F.R.S.

HAB. Kauai (Gould); Hanalei (Baldwin); Kaholuamano (Perkins).

(3) *Leptachatina antiqua* Pease.

Leptachatina antiqua Pease, P. Zool. Soc. London, 1869, p. 651 (nom. sol.);
   J. Conchyl. XVIII. (1870), p. 94; Crosse, J. Conchyl. XXIV. p. 98, pl. III. fig. 6.
Leptachatina antiqnata Pease, J. Conchyl. XVIII. (1870), p. 87 [err. typ.].

HAB. Kauai (Pease); Mana (Baldwin).

(4) *Leptachatina approximans* Ancey.

Leptachatina approximans Ancey, Naturaliste, 1897, p. 222.
HAB. Waianae, Oahu (Ancey).

(5) *Leptachatina arborea*, n. sp.

Testa ovato-oblonga, turrita, tenuis vel tenuiuscula, dextrorsa, cornea, longitudinaliter levissime striatula; anfr. 6—7, plano-convexi, ultimus ⅔ altitudinis testae
aequans; sutura bene impressa; apertura quadrato-ovata, margine dextro subincras-
satulo, columellari verticali, incrassato, reflexo, plica parva vix conspicua munito.
Alt. 8; diam. 3·6 mill. Plate XI. fig. 21.

The plica is very inconspicuous; the colour becomes lighter in adult specimens, and
then the polished, transparent gloss disappears and the shell becomes of a straw colour.
Over thirty specimens were collected by Mr Perkins; they vary slightly in shape, a
few being broader in proportion to the length, and having more inflated whorls.
Mr Baldwin sends me the following note: "It is found on the Candle-nut tree
(*Aleurites moluccana*), among the leaves of the Bird-nest fern (*Asplenium nidus*), sometimes at a height of 30 or 40 feet. All the other known species of *Leptachatina* are
terrestrial—under rocks or on dead leaves and decaying wood."

HAB. Hawaii, Kona at 4000 feet (Perkins); Olaa, Hilo (Baldwin).

(6) *Leptachatina balteata* Pease.

Leptachatina balteata Pease, P. Zool. Soc. London. 1869. p. 651 (nom. sol.):
 J. Conchyl. XVIII. (1870). p. 91 : Crosse, l. c. XXIV. (1876). p. 96, pl. IV. fig. 4.

Four, apparently immature, specimens : they approach this species very closely
save that they do not possess the colour band, and the last whorl measures just over,
rather than under, half the length of the shell. As the species is only known to me
from description and figure, I think it safer to refer them here with a query than to
describe them.

Hab. Kauai (Pease): Wahiawa (Baldwin): at 4000 feet (Perkins).

(7) *Leptachatina brevicula* Pease.

Leptachatina brevicula Pease, J. Conchyl. XVIII. (1869). p. 169.

Only known to me from the description. The specimens, while slightly larger
than the dimensions stated by Pease, agree well with the proportions given. The
plait, which he states is "*valida, fere transversa*," seems to vary much in size and
prominence.

Hab. Kauai (Pease): Kaholuamano, and at 4000 feet (Perkins).

(8) *Leptachatina (Labiella) callosa* Pfeiffer.

Achatinella (Labiella) callosa Pfeiffer. P. Zool. Soc. London. 1856 [1857] p. 334.

Only known to me from the unique type in the British Museum.

Hab. Oahu (Pfeiffer).

(9) *Leptachatina cerealis* Gould

Achatinella cerealis Gould, P. Boston Soc. II. (1847). p. 201 : U. S. Explor.
 Exped. Mollusca, pl. VII. fig. 99 ; Hartman, P. Ac. Philad. 1888, pl. I. fig. 13.

Two specimens only, which, if not this species, are probably undescribed.

Hab. Oahu, Waianae (Gould): Waianae Mts. below Kaala (Perkins).

(10) *Leptachatina chrysallis* Pfeiffer.

Achatina chrysallis Pfeiffer, P. Zool. Soc. London, 1855. p. 99.

This species has been united with *L. obtusa* Newc., by Mr Baldwin, but appears
to me to be quite distinct : the habitat he gives of 'Wahiawa to Kawailoa, Oahu'
probably really refers to *L. obtusa*.

Hab. Hawaiian Islands (Pfeiffer).

(11) *Leptachatina cingula* Mighels.

Achatinella cingula Mighels, P. Boston Soc. II. (1845), p. 21.
Leptachatina cingula Mighels, Hartman, P. Ac. Philad. 1888, pl. 1. fig. 14.
Achatinella (Leptachatina) dimidiata Pfeiffer, P. Zool. Soc. London. 1855, p. 205.

The *L. cingula* Mighels is unknown to me; I quote the following from Mr Hartman, "*Achatinella dimidiata* Pfeiffer, equals *cingula* Migh. in coll. Newcomb *ex Auct*. The figure of this shell in Chemnitz [i.e. Conch.-Cab. *Bulimacca*, pl. LXVII. figs. 5—7] does not represent the species, but equals an *Amastra*."

HAB. Oahu (Mighels, Pfeiffer); Halemano, Kawailoa Gulch (Perkins).

(12) *Leptachatina clausina* Mighels.

Bulimus clausinus Mighels, P. Boston Soc. II. (1845), p. 20.
Leptachatina clausiana (sic) Mighels, Hartman, P. Ac. Philad. 1888, p. 52.
Unknown to me.
HAB. Hawaii (Mighels).

(13) *Leptachatina columna* Ancey.

Leptachatina columna Ancey, Naturaliste. 1889, p. 266; Sykes, P. Malac. Soc. London, III. pl. XIII. fig. 18.
Near *L. chrysallis* Pfr.
HAB. Oahu (Ancey).

(14) *Leptachatina compacta* Pease.

Labiella compacta Pease. J. Conchyl. XVII. (1869), p. 172.
The specimens appear to agree with Pease's description; the species has not been figured.
HAB. Maui (Pease); E. Maui (Baldwin); Haleakala, at 5000 feet (Perkins).

(15) *Leptachatina conicoides*, sp. nov.

Testa conico-ovata, imperforata, dextrorsa, tenuiuscula, cornea, apud suturas crenulata; anfr. 6, ultimus ⅔ altitudinis testae aequans; sutura subimpressa; apertura subverticalis, sinuato-oblonga, margine dextro sub-incrassatulo, columellari reflexo, adnato, plica obliqua, parva, compressa munito, marginibus callo tenui junctis. Alt. 7·5; diam. 3·5 mill.

Plate XI. fig. 26.

A somewhat conic shell, in which, when adult, the columella plait becomes inconspicuous. One adult and three young specimens.

Hab. Molokai (Perkins).

(16) *Leptachatina convexiuscula*, sp. nov.

Testa turrita, elongata, gracilis, tenuiuscula, dextrorsa, brunneo-cornea, laevis, polita, nitida, apice obtusulo; anfr. 6½, convexi, turgiduli, ultimus $\frac{2}{3}$ altitudinis testae aequans; sutura bene impressa; apertura pyriformis, margine columellari sinuato, plica minima munito, marginibus callo tenuissimo junctis. Alt. 8; diam. 2·8 mill.

Plate XI. fig. 11.

A shell of the group of *L. exilis* Gulick, but with more inflated whorls, slightly more tapering towards the apex, and the mouth not so drawn out to the right. Three specimens.

Hab. Oahu, Waiolani (Perkins).

(17) *Leptachatina corneola* Pfeiffer.

Achatinella corneola Pfeiffer, P. Zool. Soc. London, 1845 [1846], p. 90.
Achatinella corneola Pfeiffer, Reeve, Conch. Icon. *Achatinella*, sp. 4.

Hab. Oahu? (Baldwin); Oahu, one young specimen (Hutchison).

(18) *Leptachatina coruscans* Hartman.

Leptachatina coruscans Hartman, P. Ac. Philad. 1888, p. 52, pl. 1. fig. 16.

A variable shell in thickness and colouration.

Hab. Molokai (Hartman); Kamalo (Baldwin); Kapanui, Kolamaula, and at 4000 ft. (Perkins).

(19) *Leptachatina costulata* Gulick.

Achatinella costulata Gulick, Ann. Lyc. New York, vi. (1856), p. 177, pl. vi. fig. 5; Sykes, P. Malac. Soc. London, iii. pl. xiv. fig. 4.

Newcomb united this shell with *L. semicostata* Pfeiffer, but Gulick's type is much more slender than that species, the mouth is of a different shape, and other minor differences exist, all leading me to regard it as a good species.

Hab. Oahu, Pupukea, Waimea, and Kawailoa (Gulick).

(20) *Leptachatina costulosa* Pease.

Leptachatina costulosa Pease, P. Zool. Soc. London, 1869, p. 651 (nom. sol.);
J. Conchyl. XVIII. (1870), p. 90; Crosse, l. c. XXIV. p. 96, pl. III. fig. 4.

HAB. Kauai (Pease); Waimea and Kealia (Baldwin).

(21) *Leptachatina crystallina* Gulick.

Achatinella crystallina Gulick, Ann. Lyc. New York, VI. (1856), p. 186, pl. VI.
fig. 14.

Newcomb united this species with his *L. nitida.*

HAB. Oahu, Mokuleia, Kamoo, Waialua (Gulick).

(22) *Leptachatina emerita,* sp. nov.

Testa elongata, subcylindrica, imperforata, dextrorsa, cornea vel hyalina vel
flava, tenuiuscula, sub lente longitudinaliter tenuiter striata, apice obtusulo; anfr. 6½,
plano-convexi, ultimus ⅖ altitudinis testae fere aequans; sutura impressa, marginata;
apertura ovata, margine dextro sub-incrassatulo, columellari sub-reflexo, plica parva
inconspicua ascendente munito. Alt. 8; diam. 3·5 mill.

Plate XI. fig. 10.

Variable in colour, shading from brown to a hyaline tint; adult specimens lose
their gloss and become of a straw-yellow. The columellar plait is small and inconspicuous.

HAB. Molokai, Kalamaula, and at 4000 feet (Perkins).

(23) *Leptachatina exilis* Gulick.

Achatinella exilis Gulick, Ann. Lyc. New York, VI. (1856), p. 188, pl. VI. fig. 16
[bad]; Sykes, P. Malac. Soc. London, III. pl. XIV. fig. 18.

Leptachatina cylindrata Pease, J. Conchyl. XVII. (1869), p. 168; P. Zool. Soc.
London, 1869, p. 650 (nom. sol.).

Remarkable though the distribution may be, I am unable, after a comparison of
Gulick's type with specimens of *L. cylindrata* presented by Pease to the British
Museum, to sever these two species.

HAB. Oahu, Keawaawa (Gulick).—Kauai (Pease); at 4000 feet (Perkins).

(24)　*Leptachatina extensa* Pease.

Leptachatina extensa Pease, P. Zool. Soc. London, 1869, p. 651 (nom. sol.);
J. Conchyl. XVIII. (1870), p. 92.

Four specimens, agreeing fairly well with Pease's diagnosis, are referred to this species.

HAB.　Kauai (Pease); Kaholuamano (Perkins).

(25)　*Leptachatina fumida* Gulick.

Achatinella fumida Gulick, Ann. Lyc. New York, VI. (1856), p. 181, pl. VI. fig. 9;
Sykes, P. Malac. Soc. London, III. pl. XIV. fig. 15.

Newcomb united this with his *L. vitrea*, but they appear to me to be quite distinct.

HAB.　Oahu, Waialei, Pupukea, Waimea, Kawailoa, Halemano (Gulick).

(26)　*Leptachatina fumosa* Newcomb.

Achatinella fumosa Newcomb, P. Zool. Soc. London, 1853 [1854, Nov.], p. 140,
pl. XXIII. fig. 28.

HAB.　Oahu, Manoa (Newcomb); Kawailoa Gulch (Perkins).　Only a single specimen.

(27)　*Leptachatina fusca* Newcomb.

Achatinella fusca Newcomb, Ann. Lyc. New York, VI. (1853), p. 28; P. Zool.
Soc. London, 1853 [1854, Nov.], p. 145, pl. XXIII. fig. 44.

HAB.　Oahu, Manoa (Newcomb).

(28)　*Leptachatina fuscula* Gulick.

Achatinella fuscula Gulick, Ann. Lyc. New York, VI. (1856), p. 180, pl. VI. fig. 8.

HAB.　Oahu, mountain forests of Mokuleia (Gulick).

(29) *Leptachatina glutinosa* Pfeiffer.

Achatinella (Laminella) glutinosa Pfeiffer, P. Zool. Soc. London, 1855 [1856, Feb.], p. 204.
Achatinella lacrima Gulick, Ann. Lyc. New York, VI. (1856, Dec.), p. 176, pl. VI. fig. 4; Sykes, P. Malac. Soc. London, III. pl. XIV. fig. 10.

HAB. Oahu, Lihue, Kalaikoa, Wahiawa, Halemano, Peula (Gulick); Waianae Mts., below Kaala (Perkins). Only a single specimen.

(30) *Leptachatina gracilis* Pfeiffer.

Achatinella (Achatinellastrum) gracilis Pfeiffer, P. Zool. Soc. London, 1855, p. 6, pl. XXX. fig. 22.
Achatinella elevata (Newcomb) Pfeiffer, t. c. [1856, Feb.], p. 209.
Achatinella subula Gulick, Ann. Lyc. New York, VI. [1856, Dec.], p. 191, pl. VI. fig. 19; Sykes, P. Malac. Soc. London, III. pl. XIV. fig. 16.

Gulick's species appears to be a bandless and slightly more attenuate variety.

HAB. Oahu (various authors); Palolo Valley (Gulick); Kaala (Baldwin); Waianae Mts., below Kaala, lee side (Perkins).

(31) *Leptachatina grana* Newcomb.

Achatinella grana Newcomb, Ann. Lyc. New York, VI. (1853), p. 29; P. Zool. Soc. London, 1853 [1854], p. 146, pl. XXIII. fig. 46.
Leptachatina grana Newcomb, Bland and Binney, Ann. Lyc. New York, X. p. 336 [radula].

The types of this species have met with an accident and are entirely broken. Newcomb believed that *L. granifera* Gulick [= *L. accincta* Mighels] was a synonym, but I feel doubtful of this.

HAB. Maui, Makawao (Newcomb); Haleakala, at 5000 feet (Perkins).

(32) *Leptachatina guttula* Gould.

Achatinella guttula Gould, P. Boston Soc. II. (1847), p. 201; U. S. Explor. Exped. Mollusca, pl. VII. fig. 98.
Achatinella gummea Gulick, Ann. Lyc. New York, VI. (1856), p. 182, pl. VI. fig. 10; Sykes, P. Malac. Soc. London, III. pl. XIV. fig. 1.
Achatinella fragilis Gulick, t. c. p. 183, pl. VI. fig. 11; Sykes, t. c. pl. XIV. fig. 2.

Newcomb united—I think correctly—the two Gulickian species with Gould's; Mr Baldwin has, however, in his 'Catalogue' allowed them specific rank.

HAB. Oahu, Mokuleia, Lihue, Punaluu, Hauula, Halemano (Gulick).—Maui (Gould).

(33) *Leptachatina imitatrix*, sp. nov.

Testa elongata, turrita, imperforata, dextrorsa, tenuiuscula, flavido-cornea, sub lente longitudinaliter minute striata: anfr. 6½. plano-convexi, ultimus ⅖ altitudinis testae aequans; sutura impressa; apertura elongato-ovalis, margine dextro arcuato, acuto, columellari incrassatulo, reflexo, plica minima, inconspicua munito, marginibus callo tenui junctis. Alt. 7; diam. 2·6 mill.

Plate XI. fig. 9.

Only a single specimen. It recalls *L. exilis* of Gulick, but is more conic—i.e. the upper whorls are narrower in proportion—and is of a light straw-yellow. The columellar plait is deeply-seated and inconspicuous.

Hab. Hawaii, Mauna Loa at 4000 feet (Perkins).

(34) *Leptachatina impressa* Sykes.

Leptachatina impressa Sykes, P. Malac. Soc. London, II. (1896). p. 127.

Plate XI. fig. 8.

Hab. Lanai, Mountains behind Koele (Perkins).

(35) *Leptachatina isthmica* Ancey.

Leptachatina isthmica Ancey, P. Malac. Soc. London, III. (1899). p. 270; Sykes, t. c. pl. XIII. fig. 20.

Hab. Maui, Sand Hills between East and West Maui, subfossil (Ancey).

(36) *Leptachatina konaensis*, sp. nov.

Testa elongato-ovata, imperforata, dextrorsa, tenuiuscula, cornea vel pallide cornea, longitudinaliter tenuiter striata, apice obtusulo: anfr. 6. planati, ultimus ⅖ altitudinis testae aequans; sutura impressa, marginata; apertura sinuato-ovata, columella arcuata, margine dextro intus subcalloso, columellari subreflexo, plica mediocri ascendente munito. Alt. 8; diam. 4 mill.

Plate XI. fig. 13.

Akin to *L. simplex* Pease, but is much more swollen and inflated. Six specimens.

Hab. Hawaii, Kona at 4000 feet (Perkins).

(37) *Leptachatina (Labiella) labiata* Newcomb.

Achatinella labiata Newcomb, Ann. Lyc. New York, vi. (1853). p. 27; P. Zool.
Soc. London, 1853 [1854], p. 141, pl. xxiii. fig. 33; Gwatkin. P. Ac. Philad.
1895, p. 239 [radula].
Achatinella lagena Gulick, Ann. Lyc. New York, vi. (1856). p. 175, pl. vi. fig. 3;
Sykes, P. Malac. Soc. London, iii. pl. xiv. fig. 9.
Achatinella dentata Pfeiffer, P. Zool. Soc. London, 1855. p. 7. pl. xxx. fig. 27.

I follow Newcomb in including Gulick's species, but the latter's type does not
fully shew the thickening on the columellar wall, nor the denticle on the outer lip.

HAB. Oahu, Lehui (Newcomb); Halemano, Wahiawa, Kalaikoa (Gulick);
Mount Kaala (Perkins).

(38) *Leptachatina laevis* Pease.

Leptachatina laevis Pease, P. Zool. Soc. London, 1869, p. 651 (nom. sol.);
J. Conchyl. xviii. (1870), p. 91; Crosse, l. c. xxiv. (1876), p. 96, pl. iv. fig. 6.
HAB. Kauai (Pease); Waimea (Baldwin).

(39) *Leptachatina leucochila* Gulick.

Achatinella leucochila Gulick, Ann. Lyc. New York, vi. (1856), p. 173, pl. vi.
fig. 1; Sykes, P. Malac. Soc. London, iii. pl. xiv. fig. 12.

Newcomb united this with *L. pyramis* Pfr.; I think it is quite distinct
specifically.

HAB. Kauai (Gulick).

(40) *Leptachatina lincolata* Newcomb.

Achatinella lincolata Newcomb, Ann. Lyc. New York, vi. (1853), p. 29; P. Zool.
Soc. London, 1853 [1854]. p. 140, pl. xxiii. fig. 29.

The real habitat seems somewhat uncertain; Newcomb originally gave Maui,
subsequently Hawaii, which is more probably correct.

HAB. Maui (Newcomb and Hartman).—Hawaii (Newcomb and Baldwin).

(41) *Leptachatina lucida* Pease.

Leptachatina lucida Pease, J. Conchyl. xviii. (1870), p. 93.
HAB. Kauai (Pease); Kealia (Baldwin).

(42) *Leptachatina marginata* Gulick.

Achatinella marginata Gulick, Ann. Lyc. New York, VI. (1856). p. 179. pl. VI.
fig. 7.

United by Newcomb with *L. succincta* Newc., but the present species is smaller
and more slender.

HAB. Oahu, Kalaikoa (Gulick).

(43) *Leptachatina nitida* Newcomb.

Achatinella nitida Newcomb, Ann. Lyc. New York, VI. (1853, May), p. 29;
P. Zool. Soc. London, 1853 [1854]. p. 140.
Leptachatina nitida Newcomb, Bland and Binney, Ann. Lyc. New York, X. p. 336.
pl. XV. fig. 8 [radula].

The figure given by Newcomb (P. Zool. Soc. London, 1853, pl. XXIII. fig. 30)
apparently has been taken by error from some other shell, and does not represent this
species. The form found by Mr Perkins appears to be a variety.

HAB. E. Maui (Newcomb).—Maui and Oahu (Hartman).—Oahu, Mt Kaala
(Perkins).

(44) *Leptachatina obsoleta* Pfeiffer.

Spiraxis obsoleta Pfeiffer, P. Zool. Soc. London, 1856, p. 335.

A species of the group of *L. sandwicensis*. Mr Perkins found a single young shell
on ' Haleakala at 5000 feet.' Maui. which may be the young of this species.

HAB. ? Oahu (Baldwin).

(45) *Leptachatina obtusa* (Newcomb) Pfeiffer.

Achatinella obtusa Newcomb, Pfeiffer, P. Zool. Soc. London, 1855 [1856]. p. 209.

Mr Baldwin has suggested that this species is identical with *L. chrysallis* Pfeiffer,
but I cannot agree with him.

HAB. Hawaiian Islands.—? Oahu (Baldwin).

(46) *Leptachatina octogyrata* Gulick.

Achatinella octogyrata Gulick, Ann. Lyc. New York, VI. (1856), p. 190, pl. VI.
fig. 18 : Sykes, P. Malac. Soc. London, III. pl. XIV. fig. 7.

Newcomb placed it as a synonym of *L. obclavata*, Pfr. [= *L. sandwicensis* Pfr.]

HAB. Oahu, Palolo Valley (Gulick).

(47) *Leptachatina oryza* Pfeiffer.

Achatinella (Leptachatina) oryza Pfeiffer, P. Zool. Soc. London, 1855 [1856, Feb.],
 p. 206.
Achatinella triticea Gulick, Ann. Lyc. New York, vi. (1856, Dec.), p. 184, pl. vi.
 fig. 12; Sykes, P. Malac. Soc. London, iii. pl. xiv. fig. 8.
Hab. Oahu, subfossil (Pfeiffer); Keawaawa (Gulick).

(48) *Leptachatina (Labiella) pachystoma* Pease.

Labiella pachystoma Pease, J. Conchyl. xvii. (1869), p. 171.
I am not sure if this be a true *Labiella*.
Hab. Kauai (Pease).

(49) *Leptachatina perkinsi* Sykes.

Leptachatina perkinsi Sykes, P. Malac. Soc. London, ii. (1896), p. 128.
Plate XI. fig. 30.
Hab. Lanai, Mts. behind Koele (Perkins).

(50) *Leptachatina petila* Gulick.

Achatinella petila Gulick, Ann. Lyc. New York, vi. (1856), p. 189, pl. vi. fig. 17;
 Sykes, P. Malac. Soc. London, iii. pl. xiv. fig. 14.
United by Newcomb with *L. fusca* Newc., but appears to me to be quite
distinct.
Hab. E. Oahu, Koko (Gulick).

(51) *Leptachatina pyramis* Pfeiffer.

Achatinella pyramis Pfeiffer, P. Zool. Soc. London, 1845 [1846], p. 90; Reeve,
 Conch. Icon. *Achatinella*, sp. 41 [good].
Appears from its form to be an Oahu species, and I am not sure that Pease's
localization will prove correct.
Hab. Kauai (Pease).

F. H. II. 48

(52) *Leptachatina resinula* Gulick.

Achatinella resinula Gulick, Ann. Lyc. New York, VI. (1856), p. 174, pl. VI. fig. 2 ; Sykes, P. Malac. Soc. London, III. pl. XIV. fig. 11.

HAB. Oahu, Kawailoa, Waialei, and other localities (Gulick).

(53) *Leptachatina saccula* Hartman.

Achatinella (Leptachatina) saccula Hartman. P. Ac. Philad. 1888, p. 55, pl. 1. fig. 15.

HAB. Hawaiian Islands (Hartman).

(54) *Leptachatina sandwicensis* Pfeiffer.

Achatina sandwicensis Pfeiffer, P. Zool. Soc. London, 1846 [May], p. 32.
Achatinella (Leptachatina) obelavata Pfeiffer, Op. cit. 1855 [July], p. 98.
Leptachatina octavula Paetel, Clessin, Nomenc. Helic. Viv. 1881, p. 316.

Pfeiffer placed his *Achatina sandwicensis* in the synonymy of *L. accincta* Mighels; the above identification is from an examination of Pfeiffer's types.

HAB. Oahu (Pfeiffer); Waianae Mts. (Perkins). One young specimen only.

(55) *Leptachatina saxatilis* Gulick.

Achatinella saxatilis Gulick, Ann. Lyc. New York, VI. (1856), p. 187, pl. VI. fig. 15 ; Sykes, P. Malac. Soc. London, III. pl. XIV. fig. 17.
Leptachatina saxitilus Gulick, Hartman, P. Ac. Philad. 1888, p. 55.

HAB. Oahu, Mokuleia (Gulick).

(56) *Leptachatina sculpta* Pfeiffer.

Achatina sculpta Pfeiffer, P. Zool. Soc. London, 1855 [1856], p. 211.

HAB. Oahu (Pfeiffer); (Hutchison, one specimen).

(57) *Leptachatina scutilus* Mighels.

Bulimus scutilus Mighels. P. Boston Soc. II. (1845), p. 20.

HAB. Oahu (Mighels).

(58) *Leptachatina semicostata* Pfeiffer.

Achatinella (Leptachatina) semicostata Pfeiffer, P. Zool. Soc. London. 1855 [1856, Feb.], p. 206.

Dr Hartman remarks (P. Ac. Philad. 1888, p. 55) " Dr Newcomb thinks it questionable if this species be not a synonym of *L. fusca* Newc."; it is quite distinct.

HAB. Hawaiian Islands (Pfeiffer).

(59) *Leptachatina semipicta* Sykes.

Leptachatina semipicta Sykes, P. Malac. Soc. London, II. (1896), p. 128.
Plate XI. fig. 12.
HAB. Lanai, Mts. behind Koele (Perkins).

(60) *Leptachatina simplex* Pease.

Leptachatina simplex Pease, P. Zool. Soc. London, 1869, p. 651 (nom. sol.); J. Conchyl. XVII. 1869, p. 170.

Dr Hartman notes (P. Ac. Philad. 1888, p. 55) that "Examples *L. nitida* Newc. (coll. Newc.) and *L. simplex* Pse. (coll. Pse.) are similar." There must be some error here, as specimens presented by Pease to the British Museum are quite distinct from *L. nitida* Newc.; further, Newcomb's species does not come from Hawaii.

HAB. Hawaii (Pease); Kona, at 3000—4000 feet (Perkins).

(61) *Leptachatina smithi* Sykes.

Leptachatina smithi Sykes, P. Malac. Soc. London, II. (1896), p. 128.
Plate XI. fig. 29.
HAB. Lanai, Mts. behind Koele (Perkins).

(62) *Leptachatina stiria* Gulick.

Leptachatina stiria Gulick, Ann. Lyc. New York, VI. (1856), p. 194, pl. VI. fig. 2.
HAB. Oahu, Halemano, Peula, Kawailoa (Gulick).

(63)   *Leptachatina striata* Newcomb.

Tornatellina striata Newcomb, P. Calif. Ac. II. (1861), p. 93.
From the description this appears to be close to *L. lucida* Pease.
HAB.   Kauai (Newcomb).

(64)   *Leptachatina striatella* Gulick.

Achatinella striatella Gulick, Ann. Lyc. New York, VI. (1856), p. 178, pl. VI. fig. 6 ;
    Sykes, P. Malac. Soc. London, III. pl. XIV. fig. 19.
United by Newcomb with *L. fusca* Newc., but appears to me to be distinct.
HAB.   Oahu, Keawaawa (Gulick).

(65)   *Leptachatina striatula* Gould.

Achatinella striatula Gould, P. Boston Soc. II. (1845), p. 28.
Achatinella clara Pfeiffer, P. Zool. Soc. London, 1845 [1846, Jan.], p. 90 ; Reeve,
    Conch. Icon. *Achatinella*, sp. 5.
A nice series, shewing both the form with the sutural band and the unicolorous
variety.
    HAB.   Kauai (various authors); Makaweli, Kaholuamano, Lihue, and at 4000 ft.
(Perkins).

(66)   *Leptachatina succincta* Newcomb.

Achatinella succincta Newcomb, P. Boston Soc. V. (1855), p. 220 ; Amer. J. Conch.
    II. (1866), p. 213, pl. XIII. fig. 7.
Leptachatina succinata Newcomb, Hartman, P. Ac. Philad. 1888, p. 55 (err. typ.).
HAB.   Oahu, Ewa (Newcomb) : Halemano (Perkins).   One specimen only.

(67)   *Leptachatina supracostata*, sp. nov.

    Testa elongata, turrita, imperforata, dextrorsa, tenuis, cornea, polita ; anfr. 8,
ultimus ½ longitudinis testae fere aequans, primi apud suturas subcostulati, reliqui
fere laeves ; sutura impressa, marginata, linea spirali notata ; apertura lunata, columella
sub-arcuata ; margine dextro sub-incrassatulo, columellari sub-reflexo, plica minima
oblique torta munito.   Alt. 6·3 ; diam. 2 mill.
    Plate XI. fig. 22.

Only two specimens. It belongs to the group of *L. exilis* Gulick; is inconspicuously costulate below the suture, the sculpture gradually fading out, until the last whorl hardly shews any marking beyond the lines of growth. There is a faint spiral line just below the suture.

HAB. Lanai, Mts. behind Koele (Perkins).

(68) *Leptachatina tenebrosa* Pease.

Labiella tenebrosa Pease, P. Zool. Soc. London, 1869, p. 651 (nom. sol.).
Leptachatina tenebrosa Pease, J. Conchyl. XVIII. (1870), p. 92; Crosse, l. c. XXIV.
   (1876), p. 96, pl. III. fig. 5.

HAB. Kauai (Pease); Wahiawa (Baldwin); Kaholuamano, and at 4000 feet (Perkins).

(69) *Leptachatina tenuicostata* Pease.

Leptachatina tenuicostata Pease, J. Conchyl. XVII. (1869), p. 170.
HAB. Hawaii (Pease).—Oahu (Baldwin).

I feel doubts as to the accuracy of the last locality, as Mr Baldwin marks it as a species he has not seen.

(70) *Leptachatina terebralis* Gulick.

Achatinella terebralis Gulick, Ann. Lyc. New York, VI. (1856), p. 193, pl. VI.
   fig. 21; Sykes, P. Malac. Soc. London, III. pl. XIV. fig. 3.
HAB. Oahu, Kawailoa (Gulick); Waianae Mts., below Kaala (Perkins).

(71) *Leptachatina teres* Pfeiffer.

Achatinella (Leptachatina) teres Pfeiffer, P. Zool. Soc. London, 1855 [1856],
   p. 206.
Near *L. obtusa* Newcomb.
HAB. Hawaiian Islands.

(72) *Leptachatina turgidula* Pease.

Labiella turgidula Pease, P. Zool. Soc. London, 1869, p. 651 (nom. sol.).
Leptachatina turgidula Pease, J. Conchyl. XVIII. (1870), p. 89; Crosse, l. c. XXIII.
   (1876), p. 96, pl. IV. fig. 5.
HAB. Kauai (Pease); Halemanu (Baldwin); Makaweli (Perkins). Five specimens.

(73) *Leptachatina turrita* Gulick.

Achatinella turrita Gulick, Ann. Lyc. New York, vi. (1856), p. 192, pl. vi. fig. 20 ;
 Sykes, P. Malac. Soc. London, iii. pl. xiv. fig. 6.

United by Newcomb with *L. obclavata* Pfr. [= *L. sandwicensis* Pfr.], but
L. turrita is a broader and stouter shell, of a darker colour.

Hab. Oahu, Lihue (Gulick).

(74) *Leptachatina vana* sp. nov.

Testa pyramidato-conica, dextrorsa, imperforata, tenuiuscula, brunneo-cornea, nitida,
sub lente obsolete longitudinaliter striata, sutura marginata : anfr. 8, lente accrescentes,
ultimus ⅔ altitudinis testae aequans ; apertura pyriformis, margine dextro acuto,
columellari sub-reflexo, sinuato, plica mediocri munito, marginibus callo tenuissimo
junctis. Alt. 7·8 ; diam. 3·9 mill. Plate XI. fig. 27.

Four specimens of a brownish-horny, pyramidal shell, which has no striking
characteristics.

Hab. Oahu, Mt. Kaala (Perkins).

(75) *Leptachatina vitrea* Newcomb.

Achatinella vitrea Newcomb, P. Zool. Soc. London, 1853 [1854], p. 142, pl. xxiii.
 fig. 34.

Hab. Oahu, Manoa (Newcomb).

(76) *Leptachatina vitreola* Gulick.

Achatinella vitreola Gulick, Ann. Lyc. New York, vi. (1856), p. 194, pl. vi. fig. 23.
Achatinella parvula Gulick, t. c. p. 195, pl. vi. fig. 24 ; Sykes, P. Malac. Soc.
 London, iii. pl. xiv. fig. 13.

Both were united by Newcomb with his *L. grana* ; they appear to me to be quite
distinct from that species.

Hab. Hawaiian Islands (Gulick) ; W. Maui (Baldwin for *L. parvula*).

Thaanumia Ancey.

Thaanumia omphalodes Ancey.

Thaanumia omphalodes Ancey, P. Malac. Soc. London, III. (1899), p. 269, pl. XII. fig. 8.

The type, and only. species.

HAB. Oahu, Waianae Mountains (Ancey).

Carelia H. and A. Adams.

Carelia H. and A. Adams, Genera of Recent Mollusca, II. (Feb. 1855) p. 132.

This interesting genus, confined to Kauai save for one species on the Island of Niihau, was described by Messrs H. and A. Adams, with no named type.

The anatomy has been described by Binney, P. Ac. Philad. 1876, p. 185, who points out that it agrees in general with the *Amastra* group, but differs in having a costate jaw.

(1) *Carelia bicolor* Jay.

Achatina bicolor Jay, Cat. Shells, Ed. III. (1839), p. 119, pl. VI. fig. 3.
Carelia bicolor Jay, Binney, P. Ac. Philad. 1876, p. 185, pl. VI. [anatomy].
Achatina adusta Gould, P. Boston Soc. II. (1845), p. 26.
Carelia adusta Gould, var. *angulata* Pease, J. Conchyl. XVIII. (1870), p. 403.
Achatina fuliginea Pfeiffer, P. Zool. Soc. London, 1852 [1854], p. 66; Conch.- Cab. *Achatina*, p. 267, pl. XLIII. figs. 21, 22.

HAB. Kauai (various authors).

(2) *Carelia cochlea* Reeve.

Achatina cochlea Reeve, Conch. Icon. *Achatina*, sp. 5.

The spiral sculpture is nearly obsolete in some specimens: I have seen one measuring 61 mill. in length.

HAB. Kauai (various collectors).

(3) *Carelia cumingiana* Pfeiffer.

Spiraxis cumingiana Pfeiffer, P. Zool. Soc. London, 1855, p. 106, pl. XXXII. fig. 1.

HAB. Kauai (Pfeiffer, &c.).

(4) *Carelia dolei* Ancey.

Carelia dolei Ancey, Mem. Soc. Zool. France, VI. (1893), p. 328.
HAB. Kauai, Hanalei (Ancey); Haena, subfossil (Baldwin).

(5) *Carelia glutinosa* Ancey.

Carelia glutinosa Ancey, Mem. Soc. Zool. France, VI. (1893), p. 324.
HAB. Probably Kauai. Unknown to me.

(6) *Carelia olivacea* Pease.

Carelia olivacea Pease, Amer. J. Conch. II. (1866), p. 293.
Carelia variabilis Pease, J. Conchyl. XVIII. (1870), p. 402 [with var. *viridis*];
    P. Zool. Soc. London, 1871, p. 473.
I do not quite follow why Pease described *C. variabilis*, when, in the same paper,
he stated that it and *C. olivacea* were varieties of one species.
HAB. E. Kauai (Pease).

(7) *Carelia paradoxa* Pfeiffer.

Spiraxis paradoxa Pfeiffer, P. Zool. Soc. London, 1853, p. 128.
Differs from all others known to me in its strongly granulated surface.
HAB. Kauai.

(8) *Carelia sinclairi* Ancey.

Carelia sinclairi Ancey, Mem. Soc. Zool. France, V. (1892), p. 720.
HAB. Niihau, subfossil (Ancey).

(9) *Carelia turricula* Mighels.

Achatina turricula Mighels, P. Boston Soc. II. (1845), p. 20.
Carelia turricula Mighels, Kobelt, J. B. Malak. Ges. II. (1875), p. 225, pl. VII.
    fig. 1.
Achatina obeliscus Reeve, Conch. Icon. *Achatina*, sp. 129.
Achatina newcombi Pfeiffer, P. Zool. Soc. London, 1851 [1853], p. 262.
HAB. Kauai, Hanalei (Baldwin, Perkins).

AURICULELLA Pfeiffer.

Auriculella Pfeiffer, P. Zool. Soc. London, 1855, p. 1; Mal. Blatt. II. p. 3.
The type appears to be the *Partula auricula* Fér.

(1) *Auriculella ambusta* Pease.

Auriculella ambusta Pease, J. Conchyl. XVI. (1868), p. 345.
Probably the locality suggested by Mr Baldwin is correct.
HAB. Oahu ? (Baldwin).

(2) *Auriculella auricula* Fér.

Partula auricula Férussac, Prodr. p. 66, no. 6; Voy. de Freycinet, Zool.
    p. 486.
Auricula owaihiensis Chamisso, Nov. Act. Leop. XIV. (1829), p. 639. pl. XXXVI.
    fig. 1.
Auricula sinistrorsa Chamisso, tom. cit. p. 640. pl. XXXVI. fig. 2 [spec. juv.].
Partula dumartroyi Souleyet, Rev. Zool. V. (1842), p. 102.
Bulimus armatus Mighels, P. Boston Soc. II. (1845). p. 19.

This species varies greatly in size and shape; it is generally unicolorous, varying
from nearly white, through shades of yellow and green, to brownish green; a few
specimens have a single brown band.

HAB. Oahu, Mount Tantalus, Mount Kaala, Halemano, Head of Kawailoa
Gulch (Perkins).

(3) *Auriculella brunnea* Smith.

Auriculella brunnea Smith, P. Zool. Soc. London, 1873, p. 88. pl. X. fig. 23;
    Gwatkin, P. Ac. Philad. 1895, p. 238 [radula].

Two Lanai specimens have a single darker band at the periphery; others are
unicolorous.

HAB. Molokai and Lanai (Smith); Molokai, Kalamaula, also Lanai, behind
Koele (Perkins).

(4) *Auriculella cerea* Pfeiffer.

Achatinella cerea Pfeiffer, P. Zool. Soc. London, 1855. p. 2. pl. xxx. fig. 21.

Pease has suggested (J. Conchyl. xvi. p. 343) that this is identical with *A. petitiana* Pfeiffer ; he is not improbably correct, but I have only seen the single type specimen.

Hab. Molokai (Nevill, fide specimens from Newcomb).

(5) *Auriculella chamissoi* Pfeiffer.

Achatinella (*Auriculella*) *chamissoi* Pfeiffer, P. Zool. Soc. London, 1855. p. 98.

Hab. Oahu (Baldwin).—Hawaii (fide tablet in Brit. Mus.).

(6) *Auriculella crassula* Smith.

Auriculella crassula Smith, P. Zool. Soc. London, 1873. p. 88. pl. x. fig. 22.
Auriculella ponderosa Ancey, Bull. Soc. Malac. France, vi. (1889), p. 225.

Hab. Maui, Makawao (Baldwin) ; Iao Valley, Olinda, and Haleakala at 4000 feet (Perkins).

(7) *Auriculella diaphana* Smith.

Auriculella diaphana Smith, P. Zool. Soc. London, 1873. p. 87. pl. x. fig. 25.
Auriculella patula Smith, tom. cit. p. 88, pl. x. fig. 24.

Hab. Oahu, various localities (Smith) ; Mount Tantalus, and head of Panoa Valley (Perkins).

(8) *Auriculella expansa* Pease.

Auriculella expansa Pease. J. Conchyl. xvi. (1868), p. 343. pl. xiv. fig. 8.

Hab. Hawaiian Islands (Pease).—Probably Maui (Ancey).—Kauai (Baldwin).

(9) *Auriculella lurida* Pfeiffer.

Tornatellina castanea Pfeiffer, Mon. Helic. Viv. iii. (1853), p. 524.
Achatinella (*Auriculella*) *lurida* Pfeiffer. Mon. Helic. Viv. iv. p. 570.

Pfeiffer re-named the species, apparently to avoid confusion with *Achatinella castanea* Reeve.

Hab. Maui ? (Baldwin).—Oahu, Mount Tantalus (Perkins).

(10)   *Auriculella newcombi* Pfeiffer.

Balea newcombi Pfeiffer, P. Zool. Soc. 1852 [1854], p. 67.
Achatinella obeliscus Pfeiffer, Malak. Blätt. II. (1855), p. 166.
HAB.   Molokai, Kalamaula (Perkins).

(11)   *Auriculella obliqua* Ancey.

Auriculella obliqua Ancey, Mem. Soc. Zool. France, v. (1892), p. 721; Sykes,
   P. Malac. Soc. London, III. p. 275, pl. XIII. fig. 17.
Appears to be very near *A. ambusta* Pease.
HAB.   Oahu, Waianae Mts. (Baldwin).

(12)   *Auriculella perkinsi* sp. nov.

Testa subperforata, elongato-conica, brunnea aut corneo-brunnea, linea brunnea
ad peripheriam saepe notata, nitida; anfr. 6—6½, planiusculi, ultimus ⅔ altitudinis
testae aequans; apertura auriformis, intus brunnea, margine parietali lamina obliqua
intrante, columellari lamina volvente munitis; peristoma leviter reflexum, incrassatulum.
Alt. 8; lat. 4 mill.

Plate XI. figs. 17, 18.

var. α.   Magis elongata et tenuior, flavida, peristomate albido.

I cannot identify this species with any of the numerous varieties of *A. auricula*,
and therefore describe it.   It is very variable in colour, shading from rich brown to
light yellow: when brown the band—if present—is yellowish, and conversely.   The
lip varies in colour from dark brown to white.   It is a fairly thin shell and appears to
be common.

HAB.   Oahu, ridges round Nuuanu, and Mount Tantalus (Perkins).

(13)   *Auriculella petitiana* Pfeiffer.

Tornatellina petitiana Pfeiffer, Zeitsch. Malak. IV. (1847), p. 149; Kuster, Conch.-
   Cab. *Tornatellina*, p. 153, pl. XVIII. figs. 24, 25.
HAB.   Hawaiian Islands.
See a note under *A. cerea*.

(14) *Auriculella perpusilla* Smith.

Auriculella perpusilla Smith, P. Zool. Soc. London, 1873, p. 87, pl. x. fig. 26.
HAB. Oahu, Kahalu (Smith).

(15) *Auriculella pulchra* Pease.

Auriculella pulchra Pease, J. de Conchyl. XVI. (1868), p. 346, pl. XIV. fig. 6.
Specimens presented by Pease to the British Museum under this name do not quite agree with his diagnosis, and his figure appears to have been drawn from a variety which he notes, and not the type form. I have followed the identified specimens; possibly it is a variable species, or an error may have occurred in translating his paper.
HAB. Oahu (authors); Mount Tantalus and Mount Kaala (Perkins).

(16) *Auriculella tenella* Ancey.

Auriculella tenella Ancey, Bull. Soc. Malac. France, VI. (1889), p. 232.
HAB. Oahu, Waianae (Ancey).

(17) *Auriculella tenuis* Smith.

Auriculella tenuis Smith, P. Zool. Soc. London, 1873, p. 87, pl. x. fig. 27.
Mons. Ancey has described (Bull. Soc. Malac. France, VI. p. 230) a var. *solida*.
HAB. Oahu, various localities (Smith).

(18) *Auriculella triplicata* Pease.

Auriculella triplicata Pease, J. de Conchyl. XVI. (1868), p. 346.
HAB. Maui (Hartman).—Oahu, Tantalus and Panoa (Baldwin).

(19) *Auriculella uniplicata* Pease.

Auriculella uniplicata Pease, J. de Conchyl. XVI. (1868), p. 344, pl. XIV. fig. 7.
HAB. Maui (Pease); West Maui (Baldwin).—Molokai, Kalamaula, and above Pelekunu (Perkins).

(20) *Auriculella westerlundiana* Ancey.

Auriculella westerlundiana Ancey, Bull. Soc. Malac. France, VI. (1889), p. 218 ;
Sykes, P. Malac. Soc. London, III. p. 275, pl. XIII. fig. 21.

HAB. Hawaii, Kona, and Waimea (Ancey) ; Kona at 3000 feet, and Olaa
(Perkins).

INSUFFICIENTLY KNOWN OR ERRONEOUSLY RECORDED SPECIES.

The following appear to be only manuscript names : *jucunda* Smith ; *solida*
Gulick ; *solidissima* Smith (confer Ann. Lyc. New York. X. pp. 331—2).

Bulimus pumicatus Mighels, P. Boston Soc. II. p. 19.

HAB. Oahu.

Probably this is really an *Auriculella* ; I am totally unacquainted with it.

Partula pusilla Gould, P. Boston Soc. II. p. 197 ; U. S. Explor. Exped. Mollusca,
pl. VII. fig. 90.

This species has been referred to *Auriculella*, and consequently a Hawaiian
habitat has been suggested for it ; it is really, however, a *Tornatellina* and was
described from Metia [= Mata].

FRICKELLA Pfeiffer.

Frickella Pfeiffer, P. Zool. Soc. London, 1855, p. 2 ; Mal. Blätt. II. p. 3.

Frickella amoena Pfeiffer.

Achatinella (Frickella) amoena Pfeiffer, P. Zool. Soc. London, 1855, p. 2, pl. XXX.
fig. 3.

This aberrant species appears to be a link between *Achatinella* and *Tornatellina*.
The single young shell, found by Mr Perkins, does not quite agree with the type, as
the whorls are flatter, but I am unable to sever it specifically.

HAB. Oahu, Konahuanui (Baldwin) ; Halemano (Perkins).

TORNATELLINIDAE.

TORNATELLINA Beck.

Beck (Index Moll. 1837, p. 80) proposed this name as a subgenus of *Achatina*, and placed in it four species, all of them undiagnosed. Pfeiffer in 1841 (Symb. Hist. Helic. pt. 2, p. 5) diagnosed the genus and gave (p. 130) a list of species. Previously to this Anton had proposed (1839) *Strobilus*, but in considering his claims it should be borne in mind that *Strobila* had twice previously been used in Zoology.

The Hawaiian species appear to be but little understood : the only attempt at figuring them was made by Gould, whose six figures, under one name, represent three different species.

I have endeavoured to avoid the creation of synonyms by a careful study of the descriptions and measurements given by the various authors. The habitat in the case of these very small shells is not always reliable, as they are very liable to be transported with plants, &c.

(1) *Tornatellina baldwini* Ancey.

Tornatellina baldwini Ancey, Bull. Soc. Malac. France, VI. (1889), p. 238.

HAB. Oahu, Tantalus (Ancey) ; Waianae Mts. (Perkins).—Kauai (Baldwin).

(2) *Tornatellina compacta* sp. nov.

Testa perforata, ovata, brunneo-cornea, nitidula, tenuis ; spira curta, apice obtusulo ; anfr. 5—5½, lineis incrementi bene notati, convexiusculi, regulariter et lente crescentes, sutura bene impressa ; apertura ovato-pyriformis, lamina unica pygmaea volventi interdum praedita ; peristoma simplex, margine columellari reflexo et expanso. Alt. 2·2, diam. 1·2 mm.

Plate XI. fig. 1.

A compressed, compact little form, the aperture measuring about ⅔ of the length : the whorls are somewhat convex.

HAB. Hawaii, Mauna Loa at 2000 feet, on hilo grass (Perkins).

(3) *Tornatellina confusa* sp. nov.

Pupa peponum Gould, P. Boston Soc. II. (1847), p. 197 ; U. S. Explor. Exped. Mollusca, pl. VII. figs. 104 a—c.

See for remarks under *Tornatellina peponum* Gould ; this is the edentulous form figured by him.

HAB. Kauai, Makaweli (Perkins).

(4) *Tornatellina cylindrica* sp. nov.

Testa elongata, cylindrica, cornea, perforata; anfr. 5—5½, convexiusculi, striatuli, ultimus rotundatus, ⅔ altitudinis testae aequans, sutura impressa; apertura ovata vel lunaris, lamellam in pariete gerens; columella incrassata, albida, contorta, interdum denticulo mediocri munita. Alt. 2·2, lat. vix 1 mm.

Plate XI. fig. 28.

This species may be distinguished from the true *T. peponum*, by its smaller size and more slender shape.

Hab. Oahu, Waianae Mts. (Perkins).—Kauai, Makaweli, one specimen (Perkins).

(5) *Tornatellina dentata* Pease.

Tornatellina dentata Pease, P. Zool. Soc. London, 1871, p. 460.

I identify Mr Perkins' specimens with some doubt; if not this species they belong to no other recorded Hawaiian form.

Hab. Hawaii (Pease); Puna (Baldwin); Kona at 3000 feet (Perkins).

(6) *Tornatellina curyomphala* Ancey.

Tornatellina curyomphala Ancey, Bull. Soc. Malac. France, VI. (1889), p. 239.

Not found by Mr Perkins; I have specimens from another source, without indication as to which island they come from.

Hab. W. Maui (Ancey).

(7) *Tornatellina extincta* Ancey.

Tornatellina extincta Ancey, Bull. Soc. Malac. France, VII. (1890), p. 341.

Hab. Central Maui, subfossil (Ancey).

(8) *Tornatellina gracilis* Pease.

Tornatellina gracilis Pease, P. Zool. Soc. London, 1871, p. 460.

A single shell, found by Mr Perkins, agrees well with Pease's description and measurements, save that Pease speaks of the shell being sometimes spirally sulcate, while Mr Perkins' specimen shews traces of a single spiral thread at the periphery.

Hab. Kauai (Pease).—? Hawaii, Kona at 3000 feet (Perkins).

(9)   *Tornatellina newcombi* Pfeiffer.

Tornatellina newcombi Pfeiffer, P. Zool. Soc. London, 1856, p. 335.

I am not quite clear if the localities are to be relied on: the figure given by Gould (as *T. peponum*, U. S. Explor. Exped. Moll. pl. VII. fig. 104 *c*) does not, I think, represent this species, as has been suggested.

HAB.   Maui and Oahu (Ancey).—Kauai, Oahu, and Maui (Baldwin).

(10)   *Tornatellina oblonga* Pease.

Tornatellina oblonga Pease, P. Zool. Soc. London, 1864, p. 673; Binney, Ann. Lyc. New York, XI. p. 190 [radula].
Tornatellina bacillaris Mousson, J. Conchyl. XIX. (1871), p. 16, pl. III. fig. 5.
Tornatellina oblongata Pease, Clessin, Nom. Helic. Viv. 1881, p. 343 (err. typ.).

Unknown to me as Hawaiian: it was described from the Tonga Islands.

HAB.   Oahu, Manoa (Ancey).

(11)   *Tornatellina peponum* Gould.

Pupa peponum Gould, P. Boston Soc. II. (1847), p. 197; U. S. Explor. Exped. Mollusca, pl. VII. figs. 104, 104 *d*.

Gould has undoubtedly confused three species under this name: which it therefore becomes necessary to restrict to one of his forms. I propose that it should be used for the shells figured by him as fig. 104 and fig. 104 *d*; namely the slender species with a parietal lamina and no columellar tooth: of this I have Hawaiian specimens.

The next form, that figured as figs. 104 *a—c*, has no parietal lamina, and equally no teeth on the columella; this has been found by Mr Perkins on Kauai, and is here named *T. confusa*.

The third form, figured as fig. 104 *c* (enlargement of mouth only) is a shell nearly related to *T. euryomphala*, Ancey; it is not, I think, *T. newcombi*.

HAB.   Hawaii, Hilo, also Oahu (Gould).

(12)   *Tornatellina perkinsi*, sp. nov.

Testa elongato-pyramidalis, subperforata, cornea, sub lente striatula; anfr. 7, planiusculi, lente accrescentes, sutura subimpressa; apertura ovata, lamellam sat minutam in pariete gerens; columella incrassata, albida, plicis duabus inconspicuis praedita. Alt. 3, lat. 1·2 mm.

Plate XI. fig. 14.

Its most noteworthy features are the very elongately pyramidal shape, flattened whorls, and the two inconspicuous plicae on the columella, the upper one being more deeply seated than the lower.

HAB. Kauai, Kaholuamano at 4000 feet (Perkins).

(13) *Tornatellina trochoides* sp. nov.

Testa profunde perforata, cornea, pyramidalis, pellucida, fere laevis; spira conoidea, apice acutiusculo; anfr. 7—7½, regulariter lenteque accrescentes, planiusculi, ultimus ad peripheriam subcarinatus, sutura impressa; apertura quadrato-ovata, superne angulata, laminam conspicuam validam in pariete gerens; columella brunnea, incrassata, plicis duabus munita. Alt. 4, lat. 2 mm.

Plate XI. fig. 31.

Closely related, apparently, to *T. umbilicata* Ancey; but differs from it in the relative proportions of height and breadth.

HAB. Lanai Mountains (Perkins).

(14) *Tornatellina umbilicata* Ancey.

Auriculella umbilicata Ancey, Bull. Soc. Malac. France, VI. (1889), p. 232.
Tornatellina umbilicata Ancey, Mém. Soc. Zool. France, V. p. 721.

HAB. Maui, Lahaina (Ancey).

Fam. STENOGYRIDAE.

OPEAS Albers.

Opeas Albers, Die Heliceen, 1850, p. 175; Martens, Die Heliceen, Ed. 2, 1860, p. 265 (type *Bulimus goodalli* Miller).

(1) *Opeas junceus* Gould.

Bulimus junceus Gould, P. Boston Soc. II. (1847), p. 191; U. S. Explor. Exped. Moll. pl. VII. fig. 87.

HAB. Hawaiian Islands (Gould); all the islands (Baldwin).—Oahu, Waianae Mts. (Perkins).

(2) *Opeas prestoni* Sykes.

Opeas prestoni Sykes. P. Malac. Soc. London, III. (1898), p. 73, pl. v. fig. 4.

A long series, which I am entirely unable to sever from the shell recently described by me from Ceylon. The whole genus, however, is notorious for having been discovered in widely separated localities, and probably this species has been transported with plants.

Hab.   Hawaii, Kawailoa, Mauna Loa at 1500 feet (Perkins).

(3) *Opeas pyrgiscus* Pfeiffer.

Bulimus pyrgiscus Pfeiffer, P. Zool. Soc. London, 1861, p. 24; Malak. Blätt. 1861, p. 15: Novit. Conch. III. p. 425, pl. xcvi. figs. 10—12.

This form does not seem to have been found in recent years.

Hab.   Hawaiian Islands (Pfeiffer).

Opeas striolata Pease, is recorded as Hawaiian by Nevill (Handlist Ind. Mus. pt. I. 1878, p. 166): it appears to be a manuscript name only.

Caeciliodes (Blainville) Herrmannsen.

(1) *Caecilioides baldwini* Ancey.

Caecilianella baldwini Ancey, Mem. Soc. Zool. France, v. (1892), p. 718.

I have followed Mr Smith (J. Conch. vi. pp. 341—342) as to the generic name.

Hab.   Oahu, Manoa (Ancey).

Fam. SUCCINEIDAE.

Succinea Draparnaud.

Succinea Draparnaud, Tabl. des Moll. 1801, p. 55 (first species *S. amphibia* Drap.).

The large number of unfigured species described from the Hawaiian Islands renders the identification of specimens belonging to this puzzling group by no means an easy task.

Pease proposed (J. Conchyl. xviii. 1870, p. 89) the genus *Catinella* for *Succinea rubida*; also the genus *Truella* (P. Zool. Soc. London, 1871, p. 459) for *S. elongata*. I have thought it best to leave all the forms in *Succinea*, and have listed the species in alphabetical order.

Two species collected by Mr Perkins still await identification: they are not improbably new, but so many of the described species are unknown to me that I consider it safer to leave them for the present. One is from Honolulu and Waianae Mts. in Oahu, with a dwarf variety from the mountains of Molokai at 4000 feet; the other from Kau, Hawaii; both belong to the group of *S. canella* Gould.

(1) *Succinea anrulenta* Ancey.

Succinea anrulenta Ancey, Bull. Soc. Malac. France, VI. (1889), p. 242: Sykes, P. Malac. Soc. London, III. p. 275, pl. XIII. fig. 4.

HAB. Hawaii, North Kona (Baldwin); Kona at 3000 feet (Perkins).

(2) *Succinea baldwini* Ancey.

Succinea baldwini Ancey, Bull. Soc. Malac. France, VI. (1889), p. 250.

The specimens are identified from the description alone.

HAB. Maui, Lahaina (Baldwin); Haleakala at 9000 feet (Perkins).

(3) *Succinea bicolorata* Ancey.

Succinea bicolorata Ancey, P. Malac. Soc. London, III. (1899), p. 271, pl. XII. fig. 2.

HAB. Hawaii, Waimea (Ancey).

(4) *Succinea caduca* Mighels.

Succinea caduca Mighels, P. Boston Soc. II. (1845), p. 21; Gould, U. S. Explor. Exped. Mollusca, pl. II. fig. 30.

HAB. Oahu, Waianae Mts. (Baldwin).—Molokai Mts. (Perkins).—Lanai Mts. (Perkins).

(5) *Succinea canella* Gould.

Succinea canella Gould, P. Boston Soc. II. (1847), p. 184; U. S. Explor. Exped. Mollusca, pl. II. fig. 20; Bland and Binney, Ann. Lyc. New York, X. p. 338 [jaw and radula].

Mons. Ancey has named (Bull. Soc. Malac. France, VI. pp. 245—7) varieties *crassa, obesula, mamillaris,* and *lucida.*

HAB. Maui, Lahaina (Baldwin).—Molokai and Maui (Ancey).—Maui, Haleakala at 5000 feet; Molokai Mountains (Perkins).

(6)   *Succinea casta* Ancey.

Succinea casta Ancey, P. Malac. Soc. London, III. (1899), p. 272, pl. XII, fig. 10.
HAB.   Hawaii, Olaa (Ancey).

(7)   *Succinea cepulla* Gould.

Succinea cepulla Gould, P. Boston Soc. II. (1847), p. 182 ; U. S. Explor. Exped.
Mollusca, pl. II. fig. 15.
Succinea fragilis Souleyet, Voy. Bonite, Zool. II. (1852), p. 501, pl. XXVIII. figs.
18—20 [shell and animal : nec *S. fragilis* King].
Succinea souleyeti Ancey, Bull. Soc. Malac. France, VI. (1889), p. 255.

The synonymy given here and under *S. rotundata* has been arrived at from a
careful comparison of the original descriptions and figures, and a study of the identifica-
tions made by Pfeiffer amongst the specimens in the British Museum. The shell
figured by Reeve (Conch. Icon. *Succinea*, fig. 69), purporting to be Hawaiian, under
the name of *S. fragilis* King, is not the present species.
HAB.   Hawaii (Gould, &c.).—Oahu, Tantalus. Head of Panoa Valley (Perkins).—
Molokai Mountains (Perkins).

(8)   *Succinea cinnamomea* Ancey.

Succinea cinnamomea Ancey, Bull. Soc. Malac. France. VI. (1889). p. 247.
Plate XI, fig. 32.
HAB.   Oahu, Waianae Mts. (Baldwin); Mount Kaala (Perkins).

(9)   *Succinea delicata* Ancey.

Succinea delicata Ancey, Bull. Soc. Malac. France, VI. (1889), p. 243.
HAB.   East Maui (Ancey); Kula (Baldwin).

(10)   *Succinea elongata* Pease.

Succinea elongata Pease, J. Conchyl. XVIII. (1870), p. 96.
A single specimen was found by Mr Perkins ; the species was only known to me by
the description. Authors refer to a *S. elongata* Beck, but I have failed to trace the
reference.
HAB.   Kauai (Pease); Waimea (Baldwin) ; Kaholuamano (Perkins).

(11) *Succinea explanata* Gould.

Succinea explanata Gould, U. S. Explor. Exped. Mollusca, p. 13, pl. 11. fig. 31.
HAB. Kauai (Gould); North side (Baldwin).

(12) *Succinea garrettiana* Ancey.

Succinea garrettiana Ancey, P. Malac. Soc. London, 111. (1899), p. 272. pl. xii.
  fig. 7.
HAB. Hawaii, Rainbow Falls, Hilo (Ancey).

(13) *Succinea inconspicua* Ancey.

Succinea inconspicua Ancey, P. Malac. Soc. London, 111. (1899), p. 273. pl. xii.
  fig. 9.
HAB. Hawaii, Waimea (Ancey).

(14) *Succinea konaensis* Sykes.

Succinea konaensis Sykes, P. Malac. Soc. London, 11. (1897), p. 299.
Plate XI. fig. 34.
HAB. Hawaii, Mount Kona at 4000 feet (Perkins).

(15) *Succinea lumbalis* Gould.

Succinea lumbalis Gould, P. Boston Soc. 11. (1847), p. 183; U. S. Explor.
  Exped. Moll. pl. 11. fig. 18.
HAB. Hawaii, Mauna Kea (Baldwin).—Kauai (Gould); Makaweli, and above
Waimea at 4000 feet (Perkins).

The specimens from 'above Waimea' have a more drawn out spire, but only
belong, I think, to a varietal form.

(16) *Succinea lutulenta* Ancey.

Succinea lutulenta Ancey, Bull. Soc. Malac. France, vi. (1889), p. 244.
HAB. Maui (Ancey); Ulapalakua (Baldwin).

(17) *Succinea mauiensis* Ancey.

Succinea mauiensis Ancey, Bull. Soc. Malac. France, VI. (1889), p. 248.

HAB. Maui (Ancey); Makawao (Baldwin); Haleakala at 5000 feet (Perkins).

(18) *Succinea newcombiana* Garrett.

Succinea newcombiana Garrett, P. Calif. Ac. I. (1857), p. 103.

A species closely related to *S. cepulla* and *S. rotundata*, but, if my identification of the latter is correct, this is smaller, the mouth is rounder, and the last whorl not so much inflated.

HAB. Hawaii, district of Waimea (Garrett); Kohala Mountains (Perkins).

(19) *Succinea protracta* sp. nov.

Testa tenuis, aureo-rufa vel pallide-cornea, lineis incrementi obliquis bene notata; spira producta, apice acutulo, mamillato; anfr. 3½ convexi, sutura bene impressa discreti, rapide accrescentes; apertura ovato-pyriformis, fere recta; peristoma simplex, tenue, margine columellari arcuato, haud plicato. Long. 12·7, diam. maj. 6; alt. ap. 8, lat. ap. 5 mm.

Plate XI. fig. 25.

Belongs to the group of *S. aurulenta* Ancey, but is much more elongate and slender, and the suture is more impressed. A single white specimen from ' Mauna Loa at 3500' feet I also refer to this species.

HAB. Hawaii, Kau (Perkins).

(20) *Succinea punctata* Pfeiffer.

Succinea punctata Pfeiffer, P. Zool. Soc. London, 1854 [May, 1855], p. 297; Reeve, Conch. Icon. *Succinea*, fig. 29.

The specimens from Kohala Mts. are young and have the spire slightly more depressed, but appear to belong to this form. The white spots shewn in Reeve's figure are much exaggerated; most specimens have a few blotches of a paler tint, but I have seen none regularly spotted in the way the artist has represented them.

HAB. Hawaii (Pfeiffer); Kohala Mountains, Olaa Puna, and Kona at 3000 feet (Perkins).

(21) *Succinea rotundata* Gould.

Succinea patula Mighels, P. Boston Soc. II. (1845), p. 21 [nec *S. patula* Brug., nec King].

Succinea rotundata Gould, P. Boston Soc. II. (1847), p. 182 ; U. S. Explor. Exped. Moll. pl. II. fig. 14 [shell and animal].

Succinea newcombi Pfeiffer, P. Zool. Soc. London, 1854 [May, 1855], p. 297 ; Novit. Conch. I. pl. IX. figs. 3—5 ; Reeve, Conch. Icon. *Succinea*, fig. 61.

HAB.   Oahu (Gould, &c.).—Molokai (Pfeiffer).—Hawaii, Kohala (Perkins).

See the remarks under *S. cepulla*, which species seems to be, like the present one, widely scattered over the islands.

(22) *Succinea rubella* Pease.

Succinea rubella Pease, P. Zool. Soc. London, 1871, p. 460.

The coloration of this species is very variable, and some specimens are found of a straw-colour ; it appears to be close to *S. canella* Gould.

HAB.   Lanai (Pease) ; Mountains (Perkins).

(23) *Succinea rubida* Pease.

Succinea rubida Pease, J. Conchyl. XVIII. (1870), p. 97.

HAB.   Kauai (Pease) ; North side (Baldwin).

(24) *Succinea thaanumi* Ancey.

Succinea thaanumi Ancey, P. Malac. Soc. London, III. (1899), p. 273, pl. XII. fig. 3.

HAB.   Hawaii, Olaa (Ancey).

(25) *Succinea venusta* Gould.

Succinea venusta Gould, P. Boston Soc. II. (1847), p. 186 ; U.S. Explor. Exped. Moll. pl. II. fig. 25.

HAB.   Hawaii, Mauna Kea (Baldwin) ; Kona at 3000 feet, and Olaa Puna (Perkins).

The specimens from Kona are whitish and slightly more shouldered.

(26)   *Succinea vesicalis* Gould.

Succinea vesicalis Gould. P. Boston Soc. II. (1847), p. 183; U.S. Explor. Exped. Moll. pl. II. fig. 17.

The figure given by Reeve (Conch. Icon. *Succinea*, fig. 85) does not represent this species.

Hab.   Hawaii, Mauna Kea (Gould); Kau, and Mauna Loa at 2000 ft. (Perkins).

(27)   *Succinea waianaensis* Ancey.

Succinea waianaensis Ancey. P. Malac. Soc. London, III. (1899), p. 273. pl. XII. fig. 12.

Hab.   Oahu, Waianae Mts. (Ancey).

SPECIES INSUFFICIENTLY KNOWN OR ERRONEOUSLY RECORDED.

Succinea aperta Lea. Unknown to me; it has been doubtfully referred to *S. rotundata*.

Succinea approximata Shuttleworth, Reeve, Conch. Icon. *Succinea*, fig. 27. Apparently really refers to the West Indian *S. approximans*.

Succinea pudorina Gould, Reeve, Conch. Icon. *Succinea*, figs. 43, 75.

Two apparently distinct species are figured by Reeve under this name; he records it, I think erroneously, from the 'Sandwich Islands.'

The following appear to be only manuscript names.

     S. apicalis Ancey, Makawao, Maui.

     S. tenerrima Ancey, Hilo, Hawaii.

Fam. LIMNAEIDAE.

LIMNAEA Lamarck.

Limnaea Lamarck, Mém. Soc. Hist. Nat. Paris, 1799, p. 75 (type *Helix stagnalis* L.).

The name *Limnaea* appears to have been used as early as 1791 by Poli for the animals of *Unio*, *Anodonta* and *Chama*; but his work is so peculiar, that I feel doubts if this can be regarded as preoccupying the name for a genus.

The question whether these forms—when sinistral—belong to *Physa* or *Limnaea*, has for long proved to be a difficult one. Gould, when describing his *Physa reticulata*, remarked "its form is so much that of a reversed *Limnaea* that I am almost tempted to consider it one." Pease, in his review of the Hawaiian species, stated that he was "confident," having examined the animals of several hundred specimens, that there was no species of *Physa* in the Islands. He also remarked that sinistral and dextral specimens of the same species may be found together. Prof. E. von Martens, in 1866, expressed the view, here adopted, that these Hawaiian forms were really sinistral species of *Limnaea*.

As shewing how little the literature has been studied, I may remark that as late as 1889 Mr Cooke (P. Zool. Soc. London, 1889, p. 142), when pointing out the affinities with *Limnaea* of the so-called *Physae* of Australia, notes that "Tryon mentions, but I have failed to trace on what authority, that sinistral Limnaeas occur in the Sandwich Islands."

Mighels described (P. Boston Soc. II. p. 21) two species, *producta* and *umbilicata*, as *Physae*: according to Pease the types were destroyed by fire and, as the species are unidentifiable from the diagnoses only, I have omitted them.

(1) *Limnaea aulacospira* Ancey.

Limnaea aulacospira Ancey. Naturaliste, (2) XI. (Dec. 1889), p. 290; Sykes, P. Malac. Soc. London, III. pl. XIII. fig. 19.

HAB. Maui (Ancey); Haleakala at 5000 feet, and Iao Valley (Perkins).

(2) *Limnaea binominis*, nom. nov.

Physa sandwichensis Clessin, Conchylien-Cabinet, *Physa*, p. 342, pl. XLVIII. fig. 7 (1886).
nec *Limnaea sandwichensis* Philippi.

HAB. Hawaiian Islands (Clessin).—Oahu, Mts. near Honolulu (Perkins).

(3) *Limnaea compacta* Pease.

Limnaea compacta Pease, Amer. J. Conch. VI. (1870), p. 6, pl. III. fig. 4.
Limnaea ambigua, Pease, t. c. p. 6, pl. III. fig. 5.
Physa flavida Clessin, Conchylien-Cabinet, *Physa*, p. 364, pl. LI. fig. 9 (1886).

HAB. Oahu (Pease).—(as *L. ambigua*) Kauai, Kapaa (Baldwin).—(as *L. compacta*) all the Islands (Baldwin).

(4) *Limnaea hartmanni* Clessin.

Physa hartmanni Clessin, Conchylien-Cabinet, *Physa*, p. 374, pl. LIV. fig. 9 (1886).
HAB.   Hawaii (Clessin).

Some catalogues give a *Limnaea hartmanni* of Studer and of Charpentier, but I
cannot trace a described *species* of that name : there appears to be a variety of
L. ovatus Drap bearing the name.

(5) *Limnaea moreletiana* Clessin.

Physa moreletiana Clessin, Conchylien-Cabinet, *Physa*, p. 341, pl. XLVIII, fig. 3
(1886).

Unknown to me, but from the figure I think it may be a form of *L. turgidula*
Pease.   It appears not to be the *Limnaea moreletiana* Gassies, of Adams (Gen. Rec.
Moll. II. p. 253).
HAB.   Hawaiian Islands (Clessin).

(6) *Limnaea naticoides* Clessin.

Physa naticoides Clessin, Conchylien-Cabinet, *Physa*, p. 341, pl. XLVIII, fig. 5
(1886).
HAB.   Hawaiian Islands (Clessin).

(7) *Limnaea oahuensis* Souleyet.

Limnaea oahuensis Souleyet, Voy. Bonite, Zool. II. (1852), p. 527, pl. XXIX.
figs. 38—41 [with animal] ; Reeve, Conch. Icon. *Limnaea*, sp. 90.
Limnaea affinis Souleyet, Voy. Bonite, Zool. II. p. 528, pl. XXIX. figs. 42—44.
Limnaeus sandwichensis Philippi, Arch. Naturg. II. (1845). p. 63 ; Kuster,
Conchylien-Cabinet, *Limnaea*, p. 26, pl. IV. figs. 25, 26.
Limnaea volutata, Gould, P. Boston Soc. II. (1847), p. 211 ; U. S. Explor. Exped.
Moll. pl. IX. fig. 142.

I defer to Pease's experience and unite Souleyet's two species ; though, from the
figures, I should have regarded them as distinct.   It is not the *Limnaea affinis*
of Beck.
HAB.   Oahu (Souleyet, Pease, &c.).—Oahu and Maui (Baldwin).

(8) *Limnaea peasei* Clessin.

Physa peasei Clessin, Conchylien-Cabinet, *Physa*, p. 339, pl. XLVII. fig. 8 (1886).

Judging from specimens received by the British Museum from the Morelet collection, the figure is by no means good.

HAB. Hawaiian Islands (Clessin).

(9) *Limnaea reticulata* Gould.

Physa reticulata Gould, P. Boston Soc. II. (1847), p. 214; U. S. Explor. Exped. Moll. pl. IX. fig. 140; Sowerby. Conch. Icon. *Physa*, fig. 56; Clessin, Conch.-Cab. *Physa*, p. 330. pl. XLVI. fig. 4.

Limnaea reticulata Gould, Pease, Amer. J. Conch. VI. p. 5.

Neither Sowerby nor Clessin appears to have been aware of Gould's published description of this species.

HAB. Kauai (Pease).

(10) *Limnaea rubella* Lea.

Lymnaeus rubellus Lea, Tr. Amer. Phil. Soc. IX. (1843), p. 12.

Limnaea rubella Lea, Pease, Amer. J. Conch. VI. p. 5, pl. III. figs. 1—3.

Pease was of opinion that this might prove to be a variety of *L. oahuensis* Soul.

HAB. Oahu (Lea).—Kauai (Pease); Mts. between Lihue and the sea, also Wailua river (Perkins).

(11) *Limnaea turgidula* Pease.

Limnaea turgidula Pease, Amer. J. Conch. VI. (July, 1870), p. 5, pl. III. fig. 3.

HAB. Oahu (Pease).

ERINNA A. Adams.

Erinna newcombi A. Adams.

Erinna newcombi A. Adams, P. Zool. Soc. London. 1855. p. 120; H. and A. Adams, Gen. Rec. Moll. II. p. 644. pl. CXXXVIII. fig. 9; Bland and Binney, Ann. Lyc. New York, X. p. 349 [jaw and radula]; Binney, P. Ac. Philad. 1874, p. 54. pl. V. figs. 7—10 [jaw and radula].

HAB. Kauai, Hanalei River (Baldwin, &c.). H. and A. Adams give as locality "Henata River, Kami."

See also a note on the genus by Dr Jousseaume, Rev. Mag. Zool. (3) II. (1874), p. 25.

Ancylus, Geoffroy.

Ancylus Geoffroy, Traité sommaire des coquilles,.........aux environs de Paris, 1767. p. 122 [type apparently *A. lacustris*].

Ancylus sharpi sp. nov.

Testa pygmaea, convexiuscula, hyalino-flavida ; apertura elongato-elliptica, apice obtusulo. Long. 2 : lat. 1·1 ; alt. ·8 mill.

Plate XII. figs. 14, 14 *a*.

An insignificant little form with no striking characters ; there being no other species recorded from the Islands, I venture to give these shells a name ; they are probably not adult.

Hab. Oahu, on pali, head of Nuuanu Valley (Coll. Dr B. Sharp, commisit H. A. Pilsbry).

Fam. MELANIIDAE.

Melania Lamarck.

The genus appears to have been first put forward by Lamarck in 1799 (Mém. Soc. Hist. Nat. Paris, p. 75) and to have been also characterised by him in 1801 (Syst. An. sans Vert. p. 91). In both cases the species named by him was *Melania amarula* Lam., which is therefore the type.

(1) *Melania baldwini* Ancey.

Melania baldwini Ancey, P. Malac. Soc. London, III. (July, 1899), p. 273, pl. XII. fig. 6.

Hab. Maui, Lahaina (Ancey).

(2) *Melania indefinita* Lea.

Melania indefinita Lea, P. Zool. Soc. London, 1850, p. 187 ; Reeve, Conch. Icon. *Melania*, fig. 56 ; Brot, Conch.-Cab. *Melania*, pl. XXIII. fig. 7.
Melania newcombii Lea, Pease, Amer. J. Conch. VI. p. 6 [nec Lea, fide Brot].

Hab. Oahu (Pease).

The Philippine specimens in coll. Cuming seem identical with some from Oahu, named *M. newcombii* by Pease.

(3) *Melania kauaiensis* Pease.

Melania kauaiensis Pease, Amer. J. Conch. VI. (July 1870), p. 7, pl. III. fig. 6.

HAB. Kauai (Pease).—Molokai, Pelekunu (Perkins).

Probably the species of *Melania* are scattered over the various islands and not confined to any single locality; *M. mauiensis*, for example, has been found on Maui, Molokai, Kauai, and Oahu.

(4) *Melania mauiensis* Lea.

Melania mauiensis Lea, P. Ac. Philad. VIII. (1857), p. 145; Brot, Conch.-Cab. *Melania*, p. 322, pl. XXXIII. figs. 7, 8, 8 a.

HAB. Maui (Lea).—Maui, Oahu, Kauai (Pease).—Maui, Molokai (Brot).— Molokai, in taro patches, Pelekunu (Perkins).

Large specimens were found on Molokai by Mr Perkins, exact spot not recorded, and a small race, kindly identified for me by the late Dr Brot, on Pelekunu. *Melania tahitensis* Pease MS. is stated by Brot to be a synonym. Schepman (Notes Leyden Mus. XIV. p 158) has recorded the present species from the Island of Soemba.

(5) *Melania newcombii* Lea.

Melania newcombii Lea, P. Ac. Philad. VIII. (1857), p. 145; Brot, Conch.-Cab. *Melania*, p. 213, pl. XXIV. figs. 2, 2 a.

Melania contigua Pease, Amer. J. Conch. VI. (July 1870), p. 7.

I follow Brot in uniting *M. contigua* Pease; he also places *M. oahuensis* Pease MS. and *M. paulla* Dunker MS. in the synonymy.

HAB. Oahu (Lea); In stream in mountain gulch near Honolulu (Perkins).— Kauai (Pease).

(6) *Melania verreauxiana* Lea.

Melania verrauiana (sic) Lea, P. Ac. Philad. VIII. (1857), p. 144.

Melania verreauxiana Lea, J. Ac. Philad. n. s. VI. pl. XXII. fig. 27; Brot, Conch.- Cab. *Melania*, p. 32, pl. IV. fig. 2.

Unknown to me and may not really be Hawaiian. Dr Brot considered it might be a form of *M. largillierti* Phil.

HAB. Hawaiian Islands (Lea).

Fam. PALUDESTRINIDAE.

Paludestrina D'Orbigny.

Paludestrina porrecta Mighels.

Paludina porrecta Mighels, P. Boston Soc. II. (1845). p. 22.
HAB.   Oahu (Mighels).

Fam. HELICINIDAE.

Helicina Lamarck.

In 1799 (Mém. Soc. Hist. Nat. Paris, p. 77) the genus was described but no type
or species named ; in 1801 (Syst. An. sans Vert. p. 94) the only species named was
Helicina neritella Lam., which may be taken as the type.   Lamarck refers for a figure
to Lister (Hist. Conch. fig. 59), and this illustration appears to represent a *Helicina*,
though it is hard to be certain whether it be *H. neritella* or not.

(1) *Helicina laciniosa* Mighels.

Helicina laciniosa Mighels, P. Boston Soc. II. (1845), p. 19 ; Gould, U. S. Explor.
    Exped. Moll. pl. VII. fig. 108.
A very variable shell in size and coloration ; it appears to be always more
compact and elevated than *H. sandwichiensis.*
HAB.   Oahu (Mighels).—Kauai (Baldwin).—Lanai, behind Koele ; also Kalamaula,
Molokai ; Kaala, Oahu ; and between Lihue and the sea, Kauai (Perkins).

(2) *Helicina magdalenae* Ancey.

Helicina magdalenae Ancey, Bull. Soc. Malac. France, VII. (1890), p. 342.
Helicina constricta Pfeiffer, P. Zool. Soc. London, 1848, p. 120 ; Conch.-Cab.
    Helicina, p. 22, pl. VII. fig. 37—9 [both relate to his variety only].
Pfeiffer's typical form came from ' Otaheite ' and appears to belong to a different
species to his variety, which seems to be identical with this.   Possibly forms may be
found linking *H. magdalenae* to *H. uberta.*
HAB.   Oahu, Tantalus (Ancey).

(3) *Helicina rotelloidea* Mighels.

Helicina rotelloidea Mighels, P. Boston Soc. II. (1845), p. 19; Pfeiffer, Conch.-
Cab, *Helicina*, p. 23, pl. III. fig. 40—5.
Helicina bronniana Philippi, Zeitsch. Malak. IV. (1847), p. 124.

Hab.   Oahu (Mighels, &c.).

(4) *Helicina sandwichiensis* Souleyet.

Helicina sandwichiensis Souleyet, Voy. Bonite, Zool. II. (1852), p. 529, pl. XXX,
figs. 1—5.

nec ? *H. sandwichiensis* Sowerby, Thes. Conch. III. pl. CCLXX. figs. 173—4.

A variety "β" has been recorded by Pfeiffer as from the Loyalty Islands;
probably this is an error. See Crosse, J. Conchyl. XLII. p. 405.

Hab.   Oahu, Waianae Mts. (Baldwin); at and below Kaala (Perkins).

(5) *Helicina uberta* Gould.

Helicina uberta Gould, P. Boston Soc. II. (1847), p. 202; U. S. Explor. Exped.
Moll. pl. VII. fig. 114.

Hab.   Maui and Oahu (Gould).—Oahu, below Kaala (Perkins).

SPECIES DOUBTFUL OR ERRONEOUSLY RECORDED.

Helicina antoni Pfeiffer.   Originally recorded without locality; subsequently Pfeiffer
gave the Hawaiian Islands and the Gambiers. It really appears to come from
Honduras, and the Hawaiian habitat is probably erroneous, these supposed Hawaiian
specimens belonging, as undoubtedly the Gambier Island shells do, to *H. pazi* Crosse
(J. Conchyl. XIII. p. 221, pl. VI. fig. 8).

Helicina crassilabris, Philippi. It has been suggested by Pfeiffer that this is
Hawaiian, but it really comes from Venezuela or the Caribbean Region.

Helicina fulgora Gould, originally described from Manua, Samoa Islands; it has
also been noted, but, I think, erroneously, from the Hawaiian Islands.

Helicina pisum Philippi. I think "Sandwich Is." must have been a mistake and
possibly refers to Vate or Sandwich I.: it may be a slip for Savage I., from which
specimens, inseparable from this, undoubtedly do come. This appears not to be the
H. pisum Hombr. and Jacq., which equals *H. tahitensis* Pease.

Fam. NERITIDAE.

NERITINA Lamarck.

I have not seen the first Edition of the 'Philosophie Zoologique' (1809) in which this genus is said to occur, but in the second edition (1830) the name appears in French only, with no diagnosis or named species (Vol. I. p. 321). However in his 'Hist. An. sans Vert.' it is duly given in Latin with named species (Vol. VI. pt. 2, p. 182). The first is *N. perversa* Gmel., which is the type of Montfort's *Velates* (1810) under the more correct name of *V. conoidea*, but the others belong to *Neritina* as we understand it to-day.

(1) *Neritina cariosa* Gray.

Nerita cariosa Gray, Wood, Index Test. Suppl. *Nerita* fig. 9 (1828).
Neritina sandwichensis Deshayes, An. sans Vert. Ed. 2, VIII. (1838), p. 579.
Neritina convexa Nuttall, Jay, Cat. Shells, Ed. 3, 1839, p. 66 (nom. sol.).
Neritina nuttalli Recluz, Rev. Zool. 1841, p. 276; Souleyet, Voy. Bonite, Zool. II.
    pl. XXXIV. figs. 43—46.
Neritina solidissima Sowerby, Thes. Conch. II. p. 541, pl. CXVI. fig. 573.

I have not sufficient material to determine whether the large synonymy given by Tryon (Man. Conch. X.) is fully justified. Prof. von Martens (Conch.-Cab. *Neritina*) expressed the opinion (p. 276) that *Neritina cariosa* Gray does not really belong here, but is a form of *N. mauritii*: this has been dealt with by Mr Smith (P. Zool. Soc. London, 1884, p. 275).

HAB. Hawaiian Islands (various authors).—Maui and Oahu (Baldwin).—Hawaii, Hilo (Smith).

(2) *Neritina granosa* Sowerby.

Neritina granosa Sowerby, Tank. Cat. App. p. XI. (1825); Conch. Ill. *Neritina*
    fig. 6.
Neritina papillosa Jay, Cat. Shells, Ed. 2, 1839, pl. IV. fig. 11.
Neripteron gigas Lesson, Rev. Zool. 1842, p. 187.

HAB. All the Islands (Baldwin).—Molokai, Pelekunu (Perkins).

(3) *Neritina lugubris* Philippi.

Neritina lugubris Philippi, Abbild. Conchylien, I. pt. 2, p. 20, pl. I. fig. 9 (1845).

This has been placed as a synonym of *N. cariosa*, but from the description and figure it seems to be distinct.

HAB. Hawaiian Islands (Philippi).

(4) *Neritina neglecta* Pease.

Neritina neglecta Pease, P. Zool. Soc. London, 1860, p. 435.
HAB. Hawaiian Islands (Pease).

(5) *Neritina vespertina* Nuttall.

Neritina vespertina Nuttall, Jay, Cat. Shells, Ed. 3, 1839, p. 66 (nom. sol.);
    Reeve, Conch. Icon. *Neritina*, sp. 61.
? *Neritina sandwichensis* Desh., Reeve, Conch. Icon. sp. 82 [nec Deshayes].
HAB. All the Islands (Baldwin).

In conclusion I may call attention to three species, attributed to the Islands, which do not really belong to their fauna.

Partula terrestris Pease. Apparently a manuscript name; it has appeared in Paetel's 'Catalog' and in the Mon. Helic. Viv. (Vol. VIII. p. 209) with the habitat of 'I. Sandwich.' According to Dr Hartman, it is a synonym of *P. approximans* Pease, from Raiatea.

Spiraxis sandwicensis was described by Pfeiffer (P. Zool. Soc. London, 1856, p. 335) as from the Hawaiian Islands. It appears to me to be a form of the *Bulimus lactifluus* of Pfeiffer, described from Chili, and I feel no doubt the Hawaiian habitat is erroneous.

Bulimus kanaiensis was described by Pfeiffer in the same volume (p. 332). It is probably also Chilian and very close to *Bulimus albicans* Brod.; but I am not quite sure of the identity, as the shell is slightly more succiniform.

Finally, it may be noted that a specimen of *Viviparus chinensis* Gray, doubtless imported for food, was collected by Mr Perkins at "Wailuku," Maui.

---

§ 3. Bibliographic List (arranged alphabetically).

ADAMS, A. Descriptions of two new genera and several new species of Mollusca, from the collection of Hugh Cuming, Esq. P. Zool. Soc. London, 1855 [August to December], pp. 119—124.

ADAMS, C. B. Descriptions of new species of *Partula* and *Achatinella*. Ann. Lyc. New York, v. (1852), pp. 41—44.

——. Contributions to Conchology. Vol. 1. No. 8, 1850, pp. 125—128.

ADAMS, H. & A. The Genera of recent Mollusca. London, 3 vols. 8vo, 1853—1858.

ALBERS, J. C. Die Heliceen. Berlin, 8vo, 1850.

ANCEY, C. F. Étude sur la faune malacologique des îles Sandwich. Bull. Soc. Malac. France, VI. (1889), pp. 171—258.

——. Mollusques nouveaux de l'Archipel d'Hawai, de Madagascar, et de l'Afrique équatoriale. Op. cit. VII. (1890), pp. 339—347.

——. Diagnoses de Mollusques nouveaux. Naturaliste, ser. 2, an. iii (1889), p. 266 [*Leptachatina columna*, n. sp.].

——. Descriptions de Mollusques nouveaux. T. c. pp. 290, 291 [*Limnaea aulacospira*, n. sp.].

——. Études sur la faune malacologique des îles Sandwich. Mem. Soc. Zool. France, v. (1892), pp. 708—722.

——. Études sur la faune malacologique des îles Sandwich. Op. cit. VI. (1893), pp. 321—330.

——. Descriptions de deux nouvelles espèces de Mollusques. Naturaliste, ser. 2, an. xi (1897), p. 178 [*Amastra durandi*, n. sp.].

——. Description d'un mollusque nouveau. T. c. p. 222 [*Leptachatina approximans*, n. sp.].

——. Some notes on the non-marine molluscan fauna of the Hawaiian Islands, with diagnoses of new species. P. Malac. Soc. London, III. (July, 1899), pp. 268—274, pls. XII, XIII pars.

BALDWIN, D. D. Catalogue [of] land and fresh-water shells of the Hawaiian Islands. Honolulu, 8vo, 1893, 25 pp.

——. Descriptions of new species of Achatinellidae from the Hawaiian Islands. P. Ac. Philad. 1895, pp. 214—236, pls. x, xi.

——. Descriptions of two new species of Achatinellidae from the Hawaiian Islands. Nautilus, x. (July, 1896), pp. 31, 32.

BARNACLE, H. G. Musical sounds caused by Achatinellae. J. Conch. IV. (1883), p. 118.

BECK, H. Index Molluscorum praesentis aevi musei......Christiani Frederici. Havniae, 1837, fasc. 1, 2.

BENSON, W. H. General features of Chusan, with remarks on the Flora and Fauna of that Island. Mollusca. Ann. Nat. Hist. IX. (1842), pp. 486—490.

BERGH, R. Anatomische Untersuchung des *Triboniophorus schuttei*, Kefstr., sowie von *Philomycus carolinensis* (Bosc) und *australis* (Bergh). Verh. Ges. Wien, XX. (1870), pp. 843—868, pls. XI—XIII.

BINNEY, W. G. On some of the species of naked Pneumonobranchous Mollusca of the United States. P. Boston Soc. I. (1844), pp. 51, 52 [cf. also p. 154].

——. On the anatomy and lingual dentition of *Ariolimax* and other Pulmonata. P. Ac. Philad. 1874, pp. 33—62, pls. II—XI.

——. On the genitalia, jaw, and lingual dentition of certain species of Pulmonata [with a note on the classification of the Achatinellae, by Thomas Bland]. Ann. Lyc. New York, XI. (1875), pp. 166—196, pls. XII—XVIII.

——. On the lingual dentition, jaw, and genitalia of *Carelia*, *Onchidella*, and other Pulmonata. P. Ac. Philad. 1876, pp. 13—92, pl. VI.

——. Notes on the jaw and lingual dentition of Pulmonate Mollusks. Ann. N. York Ac. III. 1884, pp. 79—136, pls. II—XVI.

BLAND, T. & BINNEY, W. G. On the lingual dentition and anatomy of *Achatinella* and other Pulmonata. Ann. Lyc. New York, X. (Nov. 1873), pp. 331—350, pls. XV, XVI.

BOETTGER, O. Die *Pupa*-Arten Oceaniens. In von Martens' Conch. Mittheil. I. (1880), pp. 45—72, pls. X—XII.

CHAMISSO, A. DE. Species novas Conchyliorum terrestrium ex insulis Sandwich dictis attulit. Acta Ac. German. XIV. (1829), pp. 639, 640, pl. XXXVI.

CLESSIN, S. Nomenclator Heliceorum Viventium. Cassel, 8vo, 1881.

COLLINGE, W. E. On a collection of Slugs from the Sandwich Islands. P. Malac. Soc. London, II. (April, 1896), pp. 46—51, figs.

——. On a further collection of Slugs from the Hawaiian (or Sandwich) Islands. Tom. cit. (Nov. 1897), pp. 293—297, figs.

——. On the anatomy and systematic position of some recent additions to the British Museum collection of Slugs. J. Malac. VII. (1900), pp. 77—85, pls. IV, V.

COOKE, A. H. On the generic position of the so-called Physae of Australia. P. Zool. Soc. London, 1889, pp. 136—143, figs.

CROSSE, H. Note complémentaire sur quelques espèces de mollusques habitant l'île Kauai (îles Hawaii). J. Conchyl. XXIV. (1876), pp. 95—99, pls. I pars, III pars, IV pars.

DESHAYES, G. P. [See FÉRUSSAC & DESHAYES.]

DIXON, G. A voyage round the world. London, 1789 [cf. p. 354].

DRAPARNAUD, J. Tableau des mollusques terrestres et fluviatiles de la France. Paris, 8vo, 1801.

——. Histoire naturelle des mollusques terrestres et fluviatiles de la France. Paris and Montpellier, 1805.

FÉRUSSAC, D. DE. Tableaux Systématiques des animaux mollusques......suivis d'un Prodrome général....... Paris, folio, 1822.

——. [Review of Swainson's description of *Achatinella*.] Bull. Sci. Nat. XVI. (1829), pp. 138—141.

—— & DESHAYES, G. P. Histoire naturelle......des mollusques terrestres et fluviatiles....... Paris, folio, 1820—1851.

FITZINGER, L. J. Verzeichniss der im Erzherzogthum Oesterreich vorkommenden Weichthiere, als Prodrom einer Fauna derselben. Beitr. Landesk. Oesterr. III. (1833), pp. 88—122.

GAIMARD, —. [See QUOY & GAIMARD.]

GARRETT, A. On new species of marine shells of the Sandwich Islands. P. Calif. Ac. I. (1854—7) (second edition, 1873), pp. 114, 115 [*Succinea newcombiana*, n. sp.].

GEOFFROY, E. L. Traité sommaire des coquilles, tant fluviatiles que terrestres, qui se trouvent aux environs de Paris. Paris, 1767, 12mo.

GOULD, A. A. [Descriptions of shells from the Sandwich Islands.] P. Boston Soc. I. (1843—4), pp. 139, 174.

——. Descriptions of land shells from the Sandwich Islands. Op. cit. II. (1845), pp. 26—28.

——. Descriptions of new shells collected by the United States Exploring Expedition belonging to the genus *Helix*. Tom. cit. (1846), pp. 171—173, 177.

——. Descriptions of new shells collected by the United States Exploring Expedition. Tom. cit. (1847), pp. 181, 182—7, 191, 197, 200—2, 211, 214.

——. United States Exploring Expedition during the years 1838, 1839, 1840, 1841, 1842, under the command of Charles Wilkes, U.S.N. Mollusca and Shells. Philad. 1852, 4to, with folio atlas (1856).

——. Descriptions of new genera and species of shells. P. Boston Soc. VIII. (Feb. 1862), pp. 280—284.

GRAY, J. E. Catalogue of the Pulmonata in the British Museum. Part I. London, 8vo, 1855.

GREEN, J. New species of *Achatina*, with remarks on the Ti, or the *Dracena terminalis*, of the Sandwich Islands. Contrib. Maclurian Lyc. I. no. 2 (July, 1827), pp. 47—50, pl. IV.

——. Remarks on the *Achatina stewartii*. T. c. no. 3 (Jan. 1829), pp. 66, 67.

GULICK, J. T. Descriptions of new species of *Achatinella* from the Hawaiian Islands. Ann. Lyc. New York, VI. pp. 173—255, pls. VI—VIII [pp. 173—230 bear date Dec. 1856, while pp. 231—255 are dated Feb. 1858: diagnoses reprinted Malak. Blätt. V. (1858), pp. 198—224].

——. On the variation of species as related to their geographical distribution, illustrated by the Achatinellinae. Nature, VI. (July 18, 1872), pp. 222—224.

——. On the classification of the Achatinellinae. P. Zool. Soc. London, 1873, pp. 89—91.

——. On diversity of evolution under one set of external conditions. J. Linn. Soc. Zool. XI. pp. 496—505.

——. Lessons in the theory of divergent evolution, drawn from the distribution of the land shells of the Sandwich Islands. P. Boston Soc. XXIV. (1890), pp. 166, 167.

—— & SMITH, E. A. Description of new species of Achatinellinae. P. Zool. Soc. London, 1873, pp. 73—89, pls. IX, X.

GWATKIN, H. M. & SUTER, H. with prefatory note by PILSBRY, H. A. Observations on the dentition of Achatinellidae. P. Ac. Philad. 1895, pp. 237—240, pl. XI pars.

HARTMAN, W. D. A bibliographic and synonymic catalogue of the genus *Auriculella*, Pfeiffer. P. Ac. Philad. 1888, pp. 14, 15.

——. A bibliographic and synonymic catalogue of the genus *Achatinella*. Tom. cit. pp. 16—56, pl. I.

——. New species of shells from the New Hebrides and Sandwich Islands. Tom. cit. pp. 250—252, pl. XII. [? 1889[1]].

HARTMANN, J. D. W. Erd- und Süsswasser-Gasteropoden. St Gall, 1840.

HASSELT, F. C. VAN. Extrait d'une lettre de F. C. van Hasselt sur les mollusques de l'île de Java, adressée au Prof. van Swinderen à Groningue. Bull. Sci. Nat. Geol. III. (1824), pp. 81—87.

HEYNEMANN, F. D. Die Zungen von *Partula* und *Achatinella*. Malak. Blätt. XIV. (1867), pp. 146—150, pl. I.

——. Die Kiefer von *Philomycus carolinensis*, Bosc, und *australis*, Bergh. Nachrbl. Deutsch. malak. Ges. III. (1871), pp. 1, 2, pl. I pars.

HYATT, A. Evolution and migration of Hawaiian land-shells. P. Amer. Ass. XLVII. (1898), pp. 357, 358.

JAY, J. C. A catalogue of the shells......in the collection of John C. Jay. Ed. 3, 1839, 4to.

JOUSSEAUME, DR. Des genres *Erinna* et *Lantzia*. Rev. Mag. Zool. (3) II. (1874), p. 25.

---

[1] The sheet is certified by a note in the volume, dated Feb. 6, 1888 (*sic*), signed E. J. Nolan, to have been presented on Oct. 23, 1888.

KEFERSTEIN, W. Ueber die Anatomie der Gattungen *Incillaria*, Benson, and *Meghimatium*, Hasselt, im Vergleich mit der von *Philomycus*, Rafinesque. Malak. Blatt. XIII. (1866), pp. 64—70, pl. I.

KOBELT. W. Conchologische Miscellen. J.B. Deutsch. malak. Ges. II. (1875), pp. 222—228, pl. VII.

——. Die geographische Verbreitung der Mollusken. III. Die Inselfaunen. Op. cit. VI. (1879), pp. 195—224.

LAMARCK, J. B. Prodrome d'une nouvelle classification des coquilles. Mém. Soc. Hist. Nat. Paris, 1799, pp. 63—91.

——. Système des animaux sans vertèbres. Paris, 1801.

——. Histoire naturelle des animaux sans vertèbres. Paris, 7 vols, 1815—1822.

LEA, I. On freshwater and land shells. Tr. Amer. Phil. Soc. IX. (1843), pp. 1—31 [*Lymnaea rubella*, n. sp., p. 12].

——. Descriptions of fifteen new species of exotic Melaniana. P. Ac. Philad. VIII. (1857), pp. 144, 145.

——. Observations on the genus *Unio*, &c. Vol. XI. 1866.

——. New Unionidae, Melaniidae, &c., chiefly of the United States. J. Ac. Philad. n. s. VI. (1867), pp. 113—187, pls. XXII—XXIV.

—— & H. C. Description of a new genus of the family Melaniana and of many new species of the genus *Melania*, chiefly collected by Hugh Cuming, Esq., during his voyage in the east, and now described. P. Zool. Soc. London, 1850, pp. 179—197 [*M. indefinita*, n. sp., p. 187].

LESSON, R. P. Description d'une espèce nouvelle de Nériptère. Rev. Zool. 1842, pp. 187, 188 [*Neripteron gigas*].

LYONS, A. B. A few Hawaiian land-shells. Hawaiian Annual, 1892, pp. 103—109, pls. I, II.

MARTENS, E. VON. Die Heliceen. Ed. 2. Leipzig, 8vo, 1860.

——. Conchological Gleanings. III. The Sandwichian species of *Limnaeus*. Ann. Nat. Hist. (3) XVII. (1866), pp. 207—210.

——. Preuss. Exped. nach Ost-Asien. Zool. Theil, Band II. Berlin, 1867.

MARTINI & CHEMNITZ. Conchylien-Cabinet. Various monographs in Editions 1 and 2.

MIGHELS, J. W. Descriptions of shells from the Sandwich Islands and other localities. P. Boston Soc. II. (1845), pp. 18—25.

MOQUIN-TANDON, A. Histoire naturelle des mollusques de France. Paris, 1855, 2 vols. and atlas.

MÖRCH, O. A. L. Quelques mots sur un arrangement des mollusques pulmonés terrestres (Géophiles, Fér.) basé sur le système naturel (suite). J. Conchyl. XIII. (1865), pp. 376—396.

MORELET, A. Testacea nova Australiae. Bull. Soc. Moselle, 1857, pp. 26—37 [*Achatinella deshayesii*, n. sp., p. 27].

——. Des genres *Erinna*, *Lithotis*, et *Lantzia*. J. Conchyl. XXIII. (1875), pp. 280, 281.

MOUSSON, A. Faune malacologique terrestre et fluviatile des îles Tonga, d'après les envois de M. le docteur Ed. Graeffe. J. Conchyl. XIX. (1871), pp. 5—34 [*Tornatellina bacillaris*, n. sp., p. 16].

MÜLLER, O. F. Vermium terrestrium et fluviatilium, seu Animalium Infusorium, Helminthicorum et Testaceorum, non Marinorum, succincta historia. Havniae et Lipsiae, 1773—1774, 2 vols.

NEWCOMB, W. Descriptions of new species of *Achatinella* from the Sandwich Islands. Ann. Lyc. New York, VI. (May, 1853), pp. 18—30.

——. Descriptions of five new species of *Achatinella*. P. Boston Soc. V. (read May, issued Sept. 1853), pp. 218—220.

——. Descriptions of seventy-nine new species of *Achatinella*, a genus of pulmoniferous mollusks, in the collection of Hugh Cuming, Esq. P. Zool. Soc. London, 1853 [Nov. 1854], pp. 128—157, pls. XXII—XXIV.

NEWCOMB, W.　Abstract of descriptions of some animals of *Achatinella*, and other remarks.　P. Zool. Soc. London, 1854 [May, 1855], pp. 310, 311.

——.　Descriptions of new species of *Achatinella*.　Ann. Lyc. New York, VI. (Oct. 1855), pp. 142—147.

——.　Synopsis of the genus *Achatinella*.　Tom. cit. (Sept. 1858), pp. 303—336.

——.　Descriptions of new species of the genera *Achatinella* and *Pupa*.　Op. cit. VII. (April, 1860), pp. 145—147.

——.　Description of new shells.　P. Calif. Ac. II. (1861), pp. 91—94.

——.　Description of new species of land-shells.　Op. cit. III. (1865), pp. 179—182.

——.　Descriptions of Achatinellae.　Amer. J. Conch. II. (July, 1866), pp. 209—217, pl. XIII.

PEASE, W. H.　Descriptions of forty-seven new species of shells from the Sandwich Islands, in the collection of Hugh Cuming, Esq.　P. Zool. Soc. London, 1860, pp. 431—438 [*Neritina neglecta*, n. sp., p. 435].

——.　Descriptions of two new species of *Helicter* (= *Achatinella*, Swainson), from the Sandwich Islands, with a history of the genus.　Op. cit. 1862, pp. 3—7.

——.　Description of new land-shells from the Islands of the Central Pacific.　Op. cit. 1864, pp. 668—676 [*Tornatellina oblonga*, n. sp., p. 673].

——.　Descriptions of new species of land-shells inhabiting Polynesia.　Amer. J. Conch. II. (Oct. 1866), pp. 289—293.

——.　Descriptions d'espèces nouvelles d'*Auriculella* provenant des îles Hawaii.　J. Conchyl. XVI. (1868), pp. 342—347.

——.　Descriptions d'espèces nouvelles du genre *Helicter*, habitant des îles Hawaii.　Op. cit. XVII. (1869), pp. 167—176.

——.　On the classification of the Helicterinae.　P. Zool. Soc. London, 1869, pp. 644—652.

——.　Observations sur les espèces de coquilles terrestres qui habitent l'île Kauai (îles Hawaii), accompagnées de descriptions d'espèces nouvelles.　J. Conchyl. XVIII. (1870), pp. 87—97.

——.　Remarques sur certaines espèces de coquilles terrestres habitant la Polynésie, et descriptions d'espèces nouvelles.　T. c. pp. 393—403.

——.　Synonymie de quelques genres et espèces de coquilles terrestres habitant la Polynésie.　Op. cit. XIX. (1871), pp. 92—97.

——.　Remarks on the species of *Melania* and *Limnaea* inhabiting the Hawaiian Islands, with descriptions of new species.　Amer. J. Conch. VI. (1871), pp. 4—7, pl. III pars.

——.　Catalogue of the land-shells inhabiting Polynesia, with remarks on their synonymy, distribution, and variation, with descriptions of new genera and species.　P. Zool. Soc. London, 1871, pp. 449—477.

PFEFFER, G.　Anatomische Untersuchung der *Achatinella vulpina*.　J.B. Deutsch. malak. Ges. IV. (1877), pp. 330—334, figs.

PFEIFFER, L.　Symbolae ad Historiam Heliceorum.　Cassel, 8vo, 1841—1846.

——.　Monographia Heliceorum Viventium.　Leipzig, 8 vols., 1848—1877.

——.　Novitates Conchologicae.　Ser. I.　Cassel, 5 vols., 1854—1879.

——.　Uebersicht der mit innern Lamellen versehenen *Helix*-Arten.　Zeitschr. Malak. II. (1845), pp. 81—87 [*Helix lamellosa*, Fér., p. 85].

——.　Ueber neue Landschnecken von Jamaika und den Sandwichinseln.　Op. cit. III. (1846), pp. 113—120.

——.　Remarks on the genus *Achatinella*, Swainson, and description of six new species from Mr Cuming's collection.　P. Zool. Soc. London, 1845 [Jan. 1846], pp. 89, 90.

PFEIFFER, L. Descriptions of thirty new species of Helicea, belonging to the collection of H. Cuming, Esq. P. Zool. Soc. London, 1846, pp. 28—34.

——. Descriptions of twenty new species of Helicea, in the collection of H. Cuming, Esq. Tom. cit. pp. 37—41.

——. Diagnosen neuer Landschnecken. Zeitschr. für Malak. IV. (1847), pp. 145—151 [*Tornatellina petitiana*, n. sp., p. 149].

——. Descriptions of twenty-nine new species of *Helicina* from the collection of H. Cuming, Esq. P. Zool. Soc. London, 1848 [April, 1849], pp. 119—125.

——. Nachträge zur L. Pfeiffer Monographia Heliceorum : zu Vol. II. Zeitschr. für Malak. VI. (1849), pp. 85—95.

——. Beschreibungen neuer Landschnecken. Zeitschr. Malak. VII. (1850), pp. 65—80 [*Helix disculus*, n. sp., p. 68 ; cf. p. 153].

——. Nothwendige Vertauschung einiger Nahmen. Zeitschr. Malak. IX. (1852), pp. 62—64.

——. Descriptions of fifty-four new species of Helicea from the collection of Hugh Cuming, Esq. P. Zool. Soc. London, 1851 [July to Dec. 1853], pp. 252—263.

——. Descriptions of sixty-six new land-shells from the collection of H. Cuming, Esq. Op. cit. 1852 [March 10 May, 1854], pp. 56—70.

——. Descriptions of nineteen new species of Helicea, from the collection of Mr Cuming. Op. cit. 1853 [1854], pp. 124—128.

——. Skizze einer Monographie der Gattung *Achatinella*. Malak. Blätt. I. (1854), pp. 112—145.

——. Versuch einer Anordnung der Heliceen nach natürlichen Gruppen. Op. cit. II. (1855), pp. 112—185.

——. Descriptions of forty-two new species of *Helix*, from the collection of H. Cuming, Esq. P. Zool. Soc. London, 1854, pp. 49—57.

——. Descriptions of fifty-seven new species of Helicea from Mr Cuming's collection. Tom. cit. 1854 [May, 1855], pp. 286—298.

——. Descriptions of twenty-seven new species of *Achatinella* from the collection of Hugh Cuming, Esq., collected by Dr Newcomb and by Mons. D. Frick, late Consul-General of France at the Sandwich Islands. Op. cit. 1855 [March], pp. 1—7, pl. XXX.

——. Descriptions of forty-seven new species of Helicea from the collection of H. Cuming, Esq. Tom. cit. [July], pp. 94—101.

——. Descriptions of nine new species of Helicea from Mr Cuming's collection. Tom. cit. [August], pp. 106—108, pl. XXXII.

——. Descriptions of twenty-three new species of *Achatinella*, collected by Mons. D. Frick in the Sandwich Islands ; from Mr Cuming's collection. Tom. cit. [Feb. 1856], pp. 202—206.

——. Descriptions of sixteen new species of *Achatinella*, from Mr Cuming's collection, collected by Dr Newcomb in the Sandwich Islands. Tom. cit. [Feb. 1856], pp. 207—210.

——. Descriptions of five new species of Terrestrial Mollusca, chiefly from the collection of H. Cuming, Esq. Tom. cit. [Feb. 1856], pp. 210, 211.

——. Weitere Beobachtungen über die Gattung *Achatinella*. Malak. Blätt. II. (1854—1855), pp. 1—7, 64—70.

——. Versuch einer Anordnung der Heliceen nach natürlichen Gruppen. Tom. cit. (1855—1856), pp. 112—185.

——. Descriptions of twenty-five new species of land-shells, from the collection of H. Cuming, Esq. P. Zool. Soc. London, 1856, pp. 32—36.

——. Descriptions of fifty-eight new species of Helicea from the collection of H. Cuming, Esq. Tom. cit. [March, 1851], pp. 324—336.

——. Ueber die in Gould's Expedition Shells beschriebenen und abgebildeten Landschnecken. Malak. Blätt. IV. (1857), pp. 29—37.

PFEIFFER, L. Neue Landschnecken. Tom. cit. pp. 85—89.

——. Diagnosen neuer Heliceen. Tom. cit. pp. 229—232.

——. Descriptions of eleven new species of land-shells from the collection of H. Cuming, Esq. P. Zool. Soc. London, 1858 [March], pp. 20—23, pl. XI.

——. Descriptions of twenty-seven new species of land-shells, from the collection of H. Cuming, Esq. Op. cit. 1859, pp. 23—29 [*Helix hystricella*, n. sp., p. 25].

——. Descriptions of eight new species of *Achatinella*, from Mr Cuming's collection. Tom. cit. pp. 30—32.

——. Descriptions of forty-seven new species of land-shells from the collection of H. Cuming, Esq. Op. cit. 1861, pp. 20—29 [*Bulimus pyrgiscus*, n. sp., p. 24].

PHILIPPI, A. R. Diagnosen einiger neuen Conchylien. Arch. Naturg. 1845, II. pp. 50—71 [*Limnaea volutata*, n. sp., p. 63].

——. Testaceorum novorum centuria (continuatio). Zeitschr. Malak. IV. (1847), pp. 113—127.

——. Abbildungen und Beschreibungen......Conchylien. Cassel, 1845—1851, 3 vols.

PILSBRY, H. A. Relations of the land-molluscan fauna of South America. P. Ac. Philad. 1899, p. 226 [reprinted Ann. Nat. Hist. IV. (1899), p. 156].

——. [See also GWATKIN, TRYON.]

QUOY, — & GAIMARD, —. Voyage autour du monde......Uranie et Physicienne....... Paris, 1824.

RECLUZ, C. A. Descriptions de quelques nouvelles espèces de Nérites vivantes. Rev. Zool. 1841, pp. 274—276.

REEVE, L. Conchologia Iconica. Monographs relating to *Achatinella, Helix*, &c.

SEMPER, C. Reisen im Archipel der Philippinen. Band III. Landmollusken. Wiesbaden, 4to, 1870—1894.

SEMPER, O. Note relative aux genres *Balea* et *Timesa*. J. Conchyl. XIV. (1866), pp. 41—45.

SENONER, —. Extrait d'une lettre par M. Senoner. Bull. Soc. Malac. Belgique, VII. pp. cxx, cxxi.

SMITH, E. A. Description of a new species of *Helix*. Ann. Nat. Hist. (4) XX. (1877 Sept.), p. 242.

——. An account of the land and fresh-water mollusca collected during the Voyage of the 'Challenger' from December 1872 to May 1876. P. Zool. Soc. London, 1884, pp. 258—281, pls. XXI, XXII.

——. [See also GULICK.]

SOULEYET, —. Voyage autour du monde......sur la corvette Bonite. Vol. II. 1852, Paris.

——. Descriptions de quelques coquilles terrestres appartenant aux genres Cyclostome, Helice, &c. Rev. Zool. 1842, pp. 101, 102.

SOWERBY, G. B. A catalogue of the shells......of the late Earl of Tankerville....... London, 8vo, 1825.

——. The Conchological illustrations. London, 8vo, 1841.

——. Thesaurus Conchyliorum. Various Monographs.

SUTER, H. [See GWATKIN.]

SWAINSON, W. The characters of *Achatinella*, a new group of terrestrial shells, with descriptions of six species. Quart. J. Sci. Lit. and Arts, I. (1828), pp. 81—86.

——. Zoological Illustrations. Ser. II. London, 1832—1833.

SYKES, E. R. Preliminary diagnoses of new species of non-marine mollusca from the Hawaiian Islands. Parts 1, 2. P. Malac. Soc. London, II. pp. 126—132 (Oct. 1896), 298, 299 (Nov. 1897).

——. Contributions towards a list of papers relating to the non-marine mollusca of the Hawaiian Islands. Hertford, 8vo, 8 pp., 1896; second edition, 1897.

——. Illustrations of, with notes on, some Hawaiian non-marine mollusca. P. Malac. Soc. London, III. (July, 1899), pp. 275, 276, pl. XIII pars, and XIV.

TRYON, G. W. Manual of Conchology. Series 2. Vols. II (1886), IX (1894). Philadelphia.

WOOD, W. Index Testaceologicus. Supplement. London, 1828.

§ 4. List of named forms which are placed in this work as varieties or synonyms.

§ 5. List of unidentified, or erroneously recorded, forms.

	PAGE
antoni Pfeiffer (*Helicina*) .	397
aperta Lea (*Succinea*)	390
apicalis Ancey (*Succinea*)	390
approximata Shuttlw. (*Succinea*)	390
crassilabris Philippi (*Helicina*) .	397
exserta Pfeiffer (*Helix*) .	293
ferruginea Baldwin (*Amastra*)	356
fornicata Gould (*Helix*) .	293
fulgora Gould (*Helicina*) .	397
kauaiensis Pfeiffer (*Bulimus*) .	399
olesonii Baldwin (*Achatinella*) .	329
pisum Philippi (*Helicina*) .	397
pudorina Gould (*Succinea*)	390
pumicatus Mighels (*Bulimus*) .	379
pusilla Gould (*Partula*) .	379
sandwicensis Pfeiffer (*Spiraxis*)	399
sandwichensis Pfeiffer (*Helix*) .	293
striolata Pease (*Opeas*) .	384
tenerrima Ancey (*Succinea*)	390
terrestris Pease (*Partula*)	399
testudinea Baldwin (*Amastra*) .	356

THE EARTHWORMS OF THE HAWAIIAN ARCHIPELAGO.

By Frank E. Beddard, M.A., F.R.S., Prosector and Vice-Secretary of the Zoological Society of London.

Some of the specimens which I have received from the Hawaiian archipelago have been already described by me[1]. Since the publication of that paper Mr Perkins has sent a second series of bottles containing a large number of fresh individuals. In the present memoir upon the earthworm fauna of this part of the world I deal with the entire series of specimens and attempt to give a complete account of all the earthworms which have been described from the Hawaiian islands, whether they are or are not contained in the collections which I have myself examined. The collections made by Mr Perkins consist of so many individuals that they probably present a very fair specimen of the Oligochaetous fauna of Hawaii. It is therefore permissible to point out what appear to me to be justifiable deductions from the material examined. The fact that the second set of specimens contained hardly anything that was not in the first set supports my contention that I have been able to study a very representative collection.

Dr Michaelsen[2] in criticising my previous paper upon this subject advanced the opinion that there are no truly indigenous worms in these oceanic islands. I myself pointed out the absence of really peculiar forms, a general feature of oceanic islands and which at least argues their comparatively short existence. Dr Michaelsen attributes the entire earthworm fauna to transference by man. A further study of the matter inclines me to agree with him.

There are many species of Lumbricidae contained in the collections which I have examined; and the list which I gave originally can be increased. But the subject does not demand, I believe, more than a mere list of the species. They are clearly to be regarded as importations due to man.

[1] On some Earthworms from the Sandwich Islands, &c. P. Z. S. 1896, pp. 194–211.
[2] Oligochaeten von den Inseln des Pacific. Zool. Jahrb. Syst. XII. 1899, p. 211.

ALLOLOBOPHORA Savigny.

(1) *Allolobophora putris* Hoffm. (This apparently is the same as Kinberg's "*Hypogacon havaicum.*")

(2) *A. foetida* (Savigny).

(3) *A. caliginosa* (Savigny).

(4) *A. nordenskioldii* Eisen.

(5) *A. limicola* Michaelsen.

(6) *A. rosea* (Savigny).

PONTOSCOLEX Schmarda.

Pontoscolex hawaiiensis Beddard, P. Z. S., 1896, p. 196.

The ubiquitous genus *Pontoscolex* occurs in the Hawaiian archipelago. I formed a new species for the representatives of this genus which were collected by Mr Perkins chiefly on the ground that the dorsal vessel was usually double for a certain extent. Dr Michaelsen has criticised my conclusion ; and it may be that he is right. In any case the genus and species which are at least hard to distinguish from the South American *P. corethrurus* occur in the most widely separated regions of the globe. Dr Eisen however[1] has lately commenced a detailed study of this genus, so that the matter of the specific identity or difference of the specimens of *Pontoscolex* found scattered over the world had better be left alone for the present.

AMYNTAS Kinberg.

The main earthworm inhabitants of this archipelago belong to the genus *Amyntas* as I think (following Michaelsen[2]) it should now be called. The much better name *Perichaeta* was used for a Dipteran genus before it was applied to an earthworm ; and it appears to me, in spite of the ingenious protest of Horst[3], that there is no way of escaping from the conclusion that a name once used cannot be resuscitated. One unrecognisable species "*Perichaeta corticis*" has been described by Kinberg[4]. The remaining species are the following :—

(1) *Amyntas peregrinus* Fletcher.

Perichaeta peregrina Fletcher, Proc. Linn. Soc. N. S. W. (2). 1. p. 969.
Perichaeta molokaiensis Beddard, P. Z. S. 1896, p. 201.
Perichaeta floweri Benham, Journ. Linn. Soc. XXVI. p. 217.

[1] Researches in American Oligochaeta, &c. P. Calif. Ac. Sci. (3). II. p. 87.
[2] Terricolen von verschiedenen Gebieten der Erde. JB. Hamb. wiss. Anst. XVI. Beiheft 2.
[3] Zool. Anzeig. 1890, p. 6.
[4] Annulata nova. Öfv. k. Svensk. Ak. Förh. 1866.

Michaelsen has suggested that the species which I described in my preliminary paper as *Perichaeta molokaiensis*, is really identical with Fletcher's *Perichaeta peregrina*, or is at least to be regarded as a "fragliches synonym." At the time that I described that species I was not so convinced as I am now of the unimportance of size as a distinguishing characteristic of species of this genus. Fletcher described his species as being 19 cm. in length, i.e. nearly double the length of the individuals of "*Perichaeta molokaiensis*" examined by myself. Moreover Fletcher has not given any details about the clitellar setae, beyond stating that they are present. This again is a matter which is apparently not of such importance as I thought: that is to say, the same species may have setae upon one, two or three or perhaps even none of the clitellar segments.

There can I think be no doubt as to the identity of Benham's "*Perichaeta floweri*" with the present species. Benham states that there are 12 setae between the male pores, which is the number given by Fletcher. I counted 15 in my specimens. But the difference is clearly negligible. Benham particularly mentions the clitellar setae as being present on segment 16 only, a state of affairs which I found also. None of the three forms comprised in the present species possess copulatory papillae; hence it is now probably to be taken as a character of this species. In this as in all other points I can detect no differences between the descriptions of Benham and of myself. We are clearly dealing with the same species, which being with very great probability—almost amounting to certainty—identical with that described as "*Perichaeta peregrina*" by Fletcher, must bear that name. I now give for the sake of others who may doubt this identification a description of my specimens.

The two individuals of this species which I have examined were 103 and 81 mm. respectively in length. The former specimen possessed 88 segments, the latter 93.

The prostomium is small and continued by grooves on to the first half of the first segment.

The dorsal pores commence upon segments 10, 11, and are visible upon the clitellum.

The clitellum occupies the usual segments, 14—16, and has few setae upon its last segment.

The male pores are not prominent and are separated by 15 setae.

I observed no genital papillae.

The first septum separates segments 4, 5; none are thickened specially.

The gizzard occupies the usual segments which are not divided by septa.

The intestine begins in 15; the caeca are present and not large.

The sperm sacs are large and occupy segments 11, 12.

The spermiducal glands extend from segments 17—21 or 22; they are broken up into lobes which have to some extent a relation to the segmentation of the gland. The curved duct communicates directly with the exterior and not through the intermediary of a terminal dilated sac.

The spermathecae are four pairs in 6—9. The pouch is sharply marked off from the long duct. The diverticulum ending in an oval dilatation is about as long as the duct part of the main pouch.

Hab.　Molokai and Mauna Loa.

(2)　*Amyntas heterochaeta* Mich.

Perichaeta heterochaeta Michaelsen, Abhandl. nat. Vereins Hamb. XI. p. 6.
P. indica Michaelsen, Arch. f. Naturg. 1892, p. 33; nec *P. indica* Horst, Vermes in Midden Sumatra, IV. p. 4.

It is rather a curious fact that the non-identity of the worm which has been called by many persons, including myself, *Perichaeta indica* with the species described by Dr Horst under that name in the memoir quoted above has not been noticed. In that memoir Dr Horst distinctly figures a terminal sac (" Kopulationstasche ") to the duct of the "prostate" gland. His figures of "Eine *Perichaeta* von Java"[1] on the other hand do not show this duct with such a terminal swelling and refer to the species which has since been called *Perichaeta indica*. It is clear that the proper name to refer to this specimen of Dr Horst must be Dr Michaelsen's name of *Perichaeta heterochaeta*, in which no such copulatory pouch is mentioned and which in other respects agrees with the worm which has everywhere received the name of *Perichaeta indica*. If it were certain, which it is not, that M. Vaillant described only one species under the name of *Perichaeta cingulata*, then that would have to be the name for the species described by Horst, for it agrees in the presence of the terminal sac where the male gland opens on to the exterior, and in some other points.

Dr Michaelsen would include as synonymous with this species my *Perichaeta nipponica*; I think that that species may be synonymous. But that is a matter which I shall enter into on a future occasion.

I now think that I was wrong in differentiating the species *P. perkinsi*. Dr Michaelsen, chiefly for the reason that he received an example from Ceylon with papillae near to the male pores, identified my species with the one called here *Amyntas heterochaeta*. I should mention however that the fact that the union of the vas deferens with the male duct is not until near to the external orifice appears to characterise at least the individual which I examined. I found in glycerine preparation of two examples of undoubted "*indica*" that there was the more general union shortly after the duct emerged from the gland.

Dr Michaelsen and I myself have called attention to the variability which this species exhibits in the presence and number of the anterior papillae and in the presence or absence of the glandular part of the male terminal apparatus. Among the very

[1] Niederl. Arch. f. Zool. IV.

numerous examples which I have examined from the Sandwich Islands I find the following state of affairs with regard to these variable structures. In 22 examples there were no glands at all; 13 had glands; in 26 specimens the glands were either small and on both sides or only present and small or well developed on one side. The proportions seem to show that the gland is disappearing. As to the head papillae—there were none at all in 24; in 14 there were three pairs on 7, 8; in 3 there were pairs on 7, 8, 9; in one there were pairs on 8, 9; in 10 there was a pair on 8; in one there were pairs on 6, 7, 8; in 52 there were various degrees of asymmetry, sometimes none being present on one side.

Examples of the species were obtained on Maui, Mauna Loa Hawaii, Halemanu Kauai, Kilauea Hawaii, Olaa Hawaii, Haleakala Maui, Iao valley Maui, Honolulu in imported earth from China.

(3)   *Amyntas hesperidum* Beddard.

Perichaeta hesperidum F. E. Beddard, Proc. Zool. Soc. 1892, p. 169.
Perichaeta sandvicensis Id., ibid. 1896, p. 203.

In my earlier paper upon this species founded upon the first gathering of Hawaiian worms I instituted a new species for some smallish worms from several islands of the archipelago. I have since re-examined the two original specimens of *Amyntas hesperidum* which I have still by me, and have compared them with some fresh individuals undoubtedly belonging to the same species but coming from Hong Kong. The result is that I have to make one or two slight corrections in my earliest account of *A. hesperidum*. I thought that I had noted a small terminal muscular bursa in that species; but on again studying the specimens and comparing them carefully with others I find that what I took to be this distinctive structure was only the commencement of the thick investing layers of the spermiducal gland duct as it traverses the body wall. There is in fact no terminal bursa. In all the specimens the spermathecae, though lying in segments 7 and 8, as I correctly stated, open backwards, i.e. in the intersegmental furrows 7, 8; 8, 9, as I also stated. I now find that this is also the case with *Perichaeta sandvicensis*. The spermathecae as a rule lie in the 7th and 8th segments but open at the posterior margins of those segments. In both worms the diverticulum is coiled and the spermiducal gland has a rather sinuous duct which passes rather forwards on its way from the gland to the exterior. In short I can detect no differences at all between the individuals which I have referred to two species. The older name must clearly have the priority and thus I must term these Sandwich Island worms *Amyntas hesperidum*, inapt though the name undoubtedly is.

This is a prevalent species in the gatherings from the islands. In my preliminary account of the Sandwich Island worms I recorded it from Mauna Loa, Lanai, Hawaii and Molokai. I have seen in the second collection forwarded to me additional specimens from Mauna Loa. I can thus improve somewhat upon my original description of this species. It is a smallish slender worm measuring up to 100 mm. in length with a diameter of about 3 mm. The number of segments is curiously constant in the individuals which I selected for counting. In two the number was 105, in a third 104. The lengths of these specimens varied somewhat—from 82 through 98 to 100 mm. It is interesting to note the constancy of the number of segments. The differences in length are of course not sufficient to be of importance and are to be accounted for by the different degrees of contraction of the individuals.

The dorsal pores commence between segments 11, 12, and are visible upon the clitellum.

The setae of a given series of segments number as follows: 1. 21; 5. 33; 12. 52; 16. 53. The setae of the first two segments are small; those of the next four are stronger, after which segments they again diminish in size.

The clitellum is sharply marked off from the segments adjacent to it and both commences and ends with its own proper segments. I could not discover any setae upon it.

Neither could I find anywhere upon the body of the worms genital papillae. The male pores are upon the usual segment; they are slightly expanded transversely and have therefore an eye-like outline. They are fairly conspicuous: in one case the end of the spermiducal gland duct was protruded for a little way. The two pores are separated by 18 setae.

There are three fairly stout intersegmental septa in front of the gizzard which are bound to each other by numerous muscular threads in the usual way; after the gizzard come two strong septa to the anterior of which the gizzard itself is bound at its posterior end by at least five muscular straps. In a specimen from Lanai septum 8/9 was present but thin.

The gizzard has the usual position that it has in this genus. The intestinal caeca are present, but are small and simple; they occupy not more than two segments.

The last of the "hearts" is in segment 13.

The two pairs of sperm sacs are in segments 11, 12; the sperm reservoirs as also usual in segments 10, 11; of the latter the anterior pair are sometimes larger than the posterior

The spermiducal glands are much incised and occupy not more than three segments. Their muscular duct is longish and curved and is unprovided with a terminal copulatory dilatation. As to the form of the spermiducal glands it is often possible to use their characteristics as apparently valid specific distinctions. But it is necessary to be accurate in their delineation and cautious as to laying too much stress upon certain features in distinguishing species, as is shown by the present species. In most of those which I examined the gland had a somewhat ear-like form, the lower margin curving upwards and forwards like the lobe of the ear. But in one example the gland was quadrangular though only occupying four segments and deeply incised in correspondence therewith.

The spermathecae are two pairs and lie in segments 7 and 8, but open on 7, 8; 8, 9. The oval pouch has a moderately long duct to which is appended a small twisted diverticulum which is sometimes longer and sometimes shorter.

We can extract from the foregoing the following definition of the species:

Size small, 100 mm.; number of segments 105. Dorsal pores from 11, 12.

Number of setae per segment up to 55. Clitellum 14–16, without setae.

No papillae. Septum 8/9 missing. Caeca present. Last heart in 13.

Sperm sacs 11, 12. Spermiducal glands not very large; duct without end sac.

Spermathecae 7, 8, with twisted tubular diverticulum.

Remarks. Dr Michaelsen has put forward grounds for believing that this species is really Dr Horst's *Amyntas annulata.* I myself suspected a possible identity. Dr Horst's original description of *annulata*, written some years ago when there was no difficulty in distinguishing from each other the very few species of the genus at that time known, was hardly complete enough for present requirements. He gives me moreover by letter good reasons for denying the identity.

Nor is there much change required to derive this form from the prevalent *Amyntas hawayanus.* I desire again to refer in connection with this possibility to a species described by myself some years since as *Perichaeta hesperidum.* That species, two individuals, arrived together with a form which I shall refer to here, *barbadensis.* The worm differs however from *barbadensis* (I have satisfied myself by a renewed examination) in a number of points. The clitellum begins and ends sharply at the boundary lines of segments 13, 14, and 16, 17, and has no setae. The spermathecae are in segments 8, 9, or at any rate open on to the boundary lines 7, 8; 8, 9. The pouches differ from those of *hawayanus* in having much coiled diverticula; the duct of the spermiducal gland thins towards its end, is longish and rather curved and has not really a terminal bulbus as I said in my original description. I have since met with other examples of the same worm from Hong Kong in a bottle containing also examples of *barbadensis.* I have examined four of these, all that I had. They have no setae on the clitellum, which commences and ends "sharply." The length is from 80—100 mm. There are no genital papillae nor are there setae upon any segment of the clitellum. The last heart is in segment 13 as is usual; the caeca are quite normal in position and present no special features of interest. The sperm sacs have a constricted-off free end as in so many forms. The spermathecae are two pairs and lie in 8 and 9 or at least open in the intersegmental grooves 7, 8; 8, 9. The diverticulum of the pouches is only of moderate length—not so long or not longer than the pouch—and is more or less closely coiled. The duct of the spermiducal gland is directed rather forward, as is so often the case in *barbadensis*, and is rather curved, especially at the end, where it is distinctly thinner. This thin termination was not observable in one individual in which the male pores had the appearance of being somewhat everted. These worms are undoubtedly my *hesperidum*.

It is interesting to find from three distant parts of the world specimens of a worm associated with a form from which they can be easily derived, by a reduction of the number of spermathecae, and by an emphasising of the slightly coiled diverticulum of the parent (?) form, by the loss of genital papillae and setae on the clitellum. The coincidences are at least noteworthy.

If we are to assume that the migration of the genus *Amyntas* from the Oriental region is due always to the interference of man, it is most peculiar that they should have been exported in lots of corresponding species. I do not however at present do more than emphasise the facts which are as has been stated above.

(4) *Amyntas hawayanus* Rosa.

Perichaeta hawayana Rosa, Ann. k. k. Hofmus. Wien, 1891, p. 396.
Perichaeta bermudensis Beddard, P. Z. S. 1892, p. 160
Perichaeta barbadensis Beddard, ibid. p. 167.
Perichaeta morrisi Beddard, ibid. p. 166.
Perichaeta mauritiana Beddard, ibid. p. 170.
Perichaeta maudhorensis Michaelsen, Arch. f. Naturg. 1892, p. 241.
Perichaeta pallida Michaelsen, ibid. p. 227.
Perichaeta amazonica Rosa, Atti R. Ac. Torino, 1894, p. 4.
Perichaeta cupulifera Fedarb, Proc. Zool. Soc. 1898, p. 445.

The collection contains a considerable number of examples of *A. hawayanus*. These show so many variations that I believe myself to be able to justify the above rather formidable list of synonyms, which are a little more extensive than the list given by Dr Michaelsen[1] in a recent paper. My original description of *Perichaeta bermudensis* was published when I was unaware of Dr Rosa's *Perichaeta hawayana*, though his publication[2] seems to antedate mine. I was led in my "Monograph of the Oligochaeta" to adhere to my species *bermudensis* on account of the fact that Dr Rosa did not mention in his description the larger size of the setae upon the anterior segments, nor the presence of setae upon the last segment of the clitellum. The number of papillae in the neighbourhood of the male pores seemed too to be different in the two series of worms from Hawaii and from the Bermudas. In the series of specimens in the

---

[1] Die Terricolen des Madagassischen Inselgebietes. Abh. senck. naturf. Ges. 1897, p. 234.
[2] Die exotischen Terricolen, &c. Ann. k. k. Hofmus. Wien 1891, p. 396.

collection made by Mr Perkins I find the following variations in structure, from what may perhaps be regarded as the typical organisation of this species. The number of the papillae in the neighbourhood of the male pores varies; I have found only one or two, three or four; Rosa says two or three. In one example however I found six of these papillae on each side.

The number of the papillae therefore does not allow of a separation of *hawayanus* and *bermudensis*.

The setae upon the anterior segments are larger than those which follow; in one example segments 4—7 were furnished with these larger setae. This was originally one of the reasons for separating *bermudensis* from *hawayanus*.

The clitellum was described by Rosa to stop short at the middle or thereabouts of the 16th segment. I have observed both this arrangement and that generally found in *bermudensis*, i.e. that the clitellum does not commence accurately at the beginning of the 14th segment while it stops short as in the typical *hawayanus*. In one example the clitellum was exactly coincident with segments 14—16.

The clitellum has usually setae upon its last segment, i.e. the 16th of the body. There are ten to fifteen of these setae. In two examples I could see no setae anywhere upon the clitellum. This seems to have been the case with the individuals examined by Rosa. *A. bermudensis* appears to always have setae upon this segment.

A feature not yet recognised in the worms which I referred to the species *bermudensis* was found in two examples of *hawayanus*. In one of them there was a pair of papillae anterior in position lying on the 7th segment near to its posterior end and the orifices of the spermathecae. In a ripe individual there was but one of these papillae, that of the right side. This fact will be seen presently to bear upon the identity of the present species with others hitherto supposed to differ specifically from it. As to internal characters the caecum of the intestine has not always the series of short outgrowths on the under surface that has been described for this species and for *bermudensis*. The spermiducal gland is generally long, occupying segments 17—22 about. Sometimes the duct is given off at the top when the gland commences in segment 17. In one specimen the gland was much abbreviated and lay only in 18, 19 on one side and 17, 18 on the other. This looks like a commencing loss of the gland which is known to occur in some other species, e.g. *Amyntas heterochaeta*. The spermathecae seem always to lie in segments 6, 7, 8. In one example they were particularly large; but, as this individual had no other features which seemed to remove it from the species, the difference in size (the pouches were as large as the gizzard) does not seem to be more than a variation to be neglected for systematic purposes.

The size of *Amyntas hawayanus* varies to some extent. The greatest and least lengths which I observed were 150 and 69 mm. The number of segments varied between 97 and 73.

Next as to the identity of *Amyntas hawayanus* with *A. barbadensis*—the original specimens of the latter were described by me from Kew Gardens, where they had been received from Barbados. In the collection of Sandwich-Islands worms before me there are a number of specimens of this species found at Honolulu in earth imported from China. Of these I have examined seven individuals.

Their size presents no difficulty for identification. They vary from 99 to 140 mm. The clitellum occupies segment 14 to about the middle of segment 16. In two specimens I found setae to be limited to the 16th segment; in the others there were setae on all the clitellar segments, but very few on 14 and 15. In one specimen the numbers on the three segments beginning with 14 are 8, 3, 13; in another 2, 2, 10. These figures agree broadly with my previous observations upon this species.

In several cases I found that the setae upon the anterior segments of the body are as in *Amyntas hawayanus* larger than those posteriorly. Segments 3—8 appeared to be thus distinguished.

The arrangement of the genital papillae is as follows:

There are either two or three in the neighbourhood of the male pores, sometimes only one. They lie either in an oblique row or in the case where there are two, one behind the other. There is in fact no difference here from the conditions which obtain in *hawayanus*.

In addition to these posteriorly placed genital papillae there are anterior papillae. One individual had a pair on segment 7; another a median papilla on the same segment. There is here again no practical difference from *A. hawayanus*.

As to internal characters the prevalent number of spermathecae is three pairs situated as are those of *A. hawayanus*. In one specimen only were there but two pairs of these organs placed in segments 6, 7. The sperm sacs often, but not always, show a constriction near to the free end, by which a small "knob" is divided off from the rest of the sac. As in *A. hawayanus* there are at least often two pairs of egg sacs in segments 13, 14. The spermiducal glands are long, occupying segments 17—21, as in *A. hawayanus*, and as in that species there is no terminal "Kopulationstasche" into which the duct of the gland opens. A character which seems to be peculiar to these worms is the occasional duplication of the dorsal vessel. I found this in four out of seven examples; the doubling commenced at the 20th segment or thereabouts, and the tube became single again about the 25th. The doubling was complete, the two halves not uniting at the septa where they traversed those plates. Of these variable characters there are only three which do not seem to occur in examples which have been referred to *A. hawayanus* and *A. bermudensis*. These are: setae upon segments 14, 15; occasional doubling of dorsal vessel; knob-like processes of sperm sacs; the presence of only two pairs of spermathecae. Were these or some of these characters united invariably together we might indeed separate the specimens as a different species: but they do not. The one example with spermathecae in 6 and 7 only had, it is true, no marked difference in size between the anterior and the posterior setae found elsewhere among the examples; and it had a median papilla upon the 7th segment, this segment being occupied in others by a pair of similar papillae. These characters however do not always coincide, for in my original paper describing the species *Perichaeta barbadensis*, I recorded the fact that in an individual with two pairs of spermathecae there was a single median papilla upon the 7th segment, as well indeed as another occupying a similar place in the 18th. To make a species of this worm we must characterise it by the two pairs of spermathecae and the median anterior instead of paired anterior papillae, as well as by the greater uniformity in the size of the body setae generally. In view of the variations which occur in individuals which no one would thus separate it seems to be unreasonable at least in the meantime to do this.

I may perhaps be allowed to point out that I was justified on the facts as originally known in making a new species for these worms. They then differed as far as was known from *Perichaeta hawayana* in having setae upon all segments of the clitellum, in possessing anterior as well as posterior genital papillae, and finally by generally having but two pairs of spermathecae.

I shall now consider the probable identity of these forms with *Perichaeta morrisi*. This species was originally distinguished from its allies by the following assemblage of anatomical features: small size, 52 mm. with however 93 segments: two pairs of spermathecae in 6, 7: median papillae in 7, 8; glandular bodies in the neighbourhood of the male pores were not seen to open by papillae; but such glands are usually associated with papillae. Rosa[1] described later examples of what appears to be the same species. His examples were larger (up to 80 mm.): setae present on all of the segments instead of only 16; clitellum occupying the whole of segments 14—16 instead of stopping towards the middle of 16; glands near male pores; in one example a median papilla upon 18 was noted; others showed variation in the anterior papillae, in one a median papilla on 6 and a pair on 7 closely approximated in the middle line with a more lateral pair on the same segment and a median papilla on 8. Another had median papillae on 6—8; a third one only on 7. Dr Rosa also found, though it was in a rudimentary condition, the septum separating segments 8, 9.

I have been able to compare with these descriptions some worms from Hong Kong among my stores of Oligochaeta. I examined many of these which I refer to the same species.

One was 93 mm. long; two papillae lay by each male pore, and on 7 there were three papillae, one median and two lateral. Setae were present on all segments of the clitellum. In other characters I found no differences from *Perichaeta morrisi* as described.

In a second individual of 80 mm. length there were also two pores in the neigh-

---

[1] Lombrichi raccolti a Sumatra &c. Ann. Mus. civ. Genova ser. 2a XVI. p. 516.

bourhood of the male pores, but side by side, instead of one in front of the other. Setae of clitellum only on 16.

In a third there was but one papilla to each male pore, and the setae on the clitellum were limited to 16. None of these latter had any anterior genital papillae. A fourth example was 94 mm. in length, with setae only on the last segment of clitellum; the sperm sacs as in the species generally in segments 11, 12, but provided with the small terminal knobs such as I have just referred to in *Perichaeta barbadensis*. The spermathecae appeared to be in 7, 8, instead of 6, 7. In all the prostates were long and had no terminal bulb. I need not enumerate in detail the various arrangements of the genital papillae in these examples from Hong Kong; but I may state generally that they varied excessively in this particular. There were often two to four papillae on the 18th segment between the male pores; it was very general to find a pair of papillae on segment 19 corresponding in position to the male pores on the foregoing segment. I observed a median papilla on each of segments 6—8 in one individual; one was anomalous by reason of the fact that the 7th segment had no less than six papillae arranged in an irregular line along the middle of that segment. The spermathecae were as a rule two pairs in 6, 7. But this character was not absolutely fixed. One specimen had an additional spermatheca in the 8th segment, but on the right side only. In this individual moreover the generally missing septum 8 was present, a circumstance which Rosa has stated for *Perichaeta morrisi*. Among the same worms there were three specimens of rather larger size. One of these was 135 mm. long and was the largest. It has 90 segments. In it the papillae were as much reduced as they ever are in this species. The larger worms with the fewer papillae and three pairs of spermathecae I consider to be the more typical *hawayanus*. In this individual (to resume) the sperm sacs had constricted apices; the spermathecae three pairs in 6—8. One papilla only to inside of male pore; setae on 16, those of segments 3—7 about enlarged. Obviously the same as this, but a little smaller, was a worm with two papillae by male pore and a single median one on 7. These larger specimens have the duct of the spermiducal gland bent into an U-shape; in the smaller and more papillated worms the duct is usually slightly curved more in the direction of a large semicircle. If we are to accept this as a species we can find no character not found in examples of the forms already treated of, except that the two pairs of spermathecae may be a segment further behind. This seems to be hardly enough as a character whereby to separate the species.

With regard to the identity of *Perichaeta mauritiana* I must chiefly refer to Dr Michaelsen. I may observe however that in the position of the spermathecae and the presence of setae upon one segment only of the clitellum, this supposed species agrees with an individual which I found myself unable to definitely distinguish from the form which I have called *Perichaeta morrisi*.

In two individuals which I refer to this species, and which are not the same that formed the basis of my original description of the species, I found the following characters. The length of one was 76 mm. There are a row of four papillae to the inside of each male pore. The setae on segments 3—7 are particularly strong. I found setae on the last segment of the clitellum, i.e. 16. The sperm sacs have a constricted extremity. The spermiducal glands are long, extending through segments 17—22. The spermathecae are in segments 6—8; on one side of the body was an additional pouch in segment 9. A second individual was much the same, but had only three papillae by each male pore and no traces of an additional spermatheca. In my original description of *Perichaeta mauritiana* I described only two pairs of spermathecae in segments 7, 8. I cannot now lay my hands on that specimen. I may however observe that a renewed examination of one of the worms which I originally referred to *barbadensis*, seems to have its two pairs of spermathecae in 7, 8, and not as I stated in 6, 7. In any case the difference does not seem to me to be important. With the present species will have to be merged I think *Perichaeta cupulifera*. There are at least no differences of great importance to distinguish that form from Dehra Dun. There is to be seen the same kind of range in the variability of the papillae which are from as small a number as only one in front of and behind each male pore to twelve or so in the neighbourhood of those pores.

Dr Michaelsen thinks that his *Perichaeta pallida* is not to be confused with *Perichaeta hawayana*. He bases this distinction upon the fact that in *pallida* the anterior setae are not much enlarged, as they are in *hawayana*, and that the male pores are more closely approximated. As to the former it would be necessary to separate from *barbadensis* one of the individuals which I have described above as belonging to that "species" if this opinion is correct. There is at least quite as much reason for uniting this species with the series concerning which the present remarks are offered, as for including Rosa's *P. amazonica*. Rosa says nothing about the increased size of the anterior setae. The fact that the clitellum has none will not I hope, after the remarks contained in the present paper, be considered as sufficient to discriminate the species.

In Dr Michaelsen's description of *Perichaeta maudhorensis* there are no salient points which serve to discriminate it from the present species. It has larger setae on segments 2—9: the caeca have the crenated appearance below that is at least often found in *hawayana*. There is one papilla near each male pore: the three spermathecae occupy the same segments: the spermiducal glands are without the terminal sac. The sperm sacs are divided (as in some individuals of the present species) by a constriction. There is in short nothing of importance in the description which warrants a separation.

(5) *Amyntas schmardae* Horst.

Megascolex schmardae Horst, Notes Leyd. Mus. 1883, p. 194.
Perichaeta trityphla Beddard, P. Zool. Soc. 1896, p. 205.
Perichaeta vesiculata Goto and Hatai, Annot. Zool. Japon. III. p. 21.

It is rather curious that in an appendix to my account of the earthworms of the
Sandwich Islands I should have described from Barbados a species which I regarded
as new and described as *Perichaeta trityphla*.   Curious since I have subsequently found
many specimens of this worm in gatherings from Honolulu at 2000 feet of altitude and
also in earth imported from Hong Kong.   I think that Michaelsen[1] is probably right in
identifying my *trityphla* with the long known species *schmardae*.   But at the same time
it must be borne in mind that the condition of the caeca, which I thought to be distinctive of *trityphla*, has not been described in the original specimens of *schmardae*.
These caeca, I may say, vary in number from three to six on each side, and the two
sides are not always symmetrical in this respect.   The occurrence of such caeca seems
to mark the species as a native of Japan—one of the most prominent characteristics of
the species of that island being the frequent complication of the caeca.   I think that
there can be no doubt that *Perichaeta vesiculata* of Goto and Hatai is the same species.
There is positively no feature in their rather short description of the so-called *vesiculata*,
which does not fit in with the facts observed in the anatomy of *schmardae* and *trityphla*.

[1] Oligochaeten von den Inseln des Pacific. Zool. Jahrb. Syst. XII. p. 224.

ENTOZOA[1].

By Arthur E. Shipley, Fellow and Tutor of Christ's College, Cambridge, and University Lecturer in the Advanced Morphology of the Invertebrata.

THE Hawaiian Archipelago or the Sandwich Islands are separated by some 2350 miles from the mainland and by about the same distance from any other group of inhabited islands. Hence, as might be expected, their fauna is highly specialized, and although we know very little about the Entozoa of this island group, two at least of the forms described, living within the bodies of birds characteristic of the country, are new and up to the present time have not been recorded from elsewhere.

But although the Hawaiian Archipelago is so far from other lands it is a much frequented spot. Since the Spaniards first found it, before the visits in 1778 of Captain Cook, it has by degrees become one of the meeting places of the world. Its position "at the cross-roads of the North Pacific" on the line of the great marine trade-routes between Northern America, Japan, China and Australia has attracted to its harbours men of all nations, so that, like Singapore, it has a most mixed population. And man has brought not only his own parasites with him but has imported his domesticated cattle with their entozoa.

What little I have been able to find out about the human parasites in the Sandwich Islands I owe to a paper by Dr A. Lutz[2], and as the parasites he mentions are well known and almost cosmopolitan I have said little about them, still they should be mentioned in a Fauna Hawaiiensis, for though probably the majority of them have been introduced and may not be endemic, this is not certainly the case and is susceptible of no proof.

---

[1] I have much pleasure in thanking Prof. E. Ray Lankester for permission to reprint matter and figures from the Quarterly Journal of Microscopical Science; Mr Shipley has availed himself of this permission in his communication. EDITOR.

[2] Centrbl. Bakter. XIII. 1893, p. 126.

I. NEMATODA.

The following Nematodes are recorded by Dr Lutz :—

(1) *Ascaris lumbricoides* L.

This was found with *Ancylostoma duodenale* but also—and this was especially the case in children—alone.   It is said to be very common among the poorer inhabitants.

(2) *Trichocephalus dispar* Rud.

This was determined only from eggs which passed from the intestine.   It appeared to be widely distributed but not in very great numbers.

(3) *Oxyurus vermicularis* L.

Observed in one white family who had become infected through an adopted child. It is a very common parasite among the poor.

(4) *Strongyloides intestinalis* Bavery.

This parasite was observed in company with *Ancylostoma duodenale*, but in the larval state.   It apparently produced no special symptoms, and Dr Lutz repeats his doubts as to its being the cause of the so-called Cochinchina-diarrhoea.

(5) *Ancylostoma duodenale* Dubini.

The anaemia and accompanying troubles due to the presence of this parasite are very common in the Hawaiian Archipelago.   Dr Lutz found it only amongst the Portuguese, employed for the most part in the sugar plantations and careless about their supply of drinking water.   These Portuguese came not from Europe but from Madeira and the Azores and it seems probable that they brought their parasites with them.   Dr L. F. Alvarez of the " Hospital for the Treatment of Leprosy" tells me that this entozoon is very common amongst the Portuguese labourers of the city of Hilo and its neighbourhood in the Island of Hawaii.   It produces a very severe form of anaemia.

(6) *Sclerostomum armatum* Dies.

This parasite was found in the Horse and at least in one district was the cause of the death of the host.

(7) *Filaria immitis* Leidy.

Found in the heart of dogs.   It is widely spread through the Pacific Islands[1].

[1] Shipley, P, Cambridge Soc. VIII. 1892—3, p. 211.

II. PLATYHELMINTHES.

A. TREMATODA.

(1) *Distoma clavatum* Rud.

Found in the stomach of the *Coryphaena hippuris*. Several other unidentified species of Trematode were met with in other fish.

(2) *Distoma hepaticum*[1] L.

The presence of this parasite caused a veritable epizootic amongst the cattle on many of the islands of the Hawaiian Archipelago. Horses, though to a less extent, and wild swine are also said to have suffered. Dr Lutz was successful in cultivating the embryos of the Fluke, and he also succeeded in finding and observing the development of the Redia in the fresh-water snail *Limnaeus pereger*. In his second paper[1] Dr Lutz gives the opinions of certain distinguished conchologists on the species of water snails which he found infected with the larvae of *Distoma*. These Molluscs seem to afford matter for a considerable amount of disagreement and the reader is referred to this paper for the details of the subject. Dr Alvarez, to whose kindness I am indebted for several details in this paper, tells me that this Fluke sometimes attacks man.

B. CESTODA.

(1) *Drepanidotaenia hemignathi*[2] Shipley.

Plate XIII.

The specimens of this tapeworm, of which I received but ten, are all small; they vary in length from 10 mm. to 22 mm. The head is very small; immediately behind it, there being practically no neck, the body begins to broaden out, and in some specimens the proglottides attain a width of 2 mm. The segmentation of the body commences immediately behind the head, and is very well marked a little further back. The posterior border of each segment overlaps the succeeding one with a prominent edge or rim; this is well shown in longitudinal section (fig. 6). The number of segments varies from some fifty to sixty to over a hundred. The measurements given above are about the average, but, as is well known, tapeworms are extremely extensible animals, and this to a great extent diminishes the value of figures quoted in reference to their size. In some of my specimens the body is stretched, and the length of the segments equals one-half or even two-thirds of their breadth, but in the commoner forms the segments are very short and broad, sometimes eight or ten times as broad as long. They are flattened, as is seen in transverse section, and sometimes, especially towards

---

[1] Centrbl. Bakter. IX. 1892, p. 783, and XIII. 1893, p. 320.

[2] The description of this species is reprinted (with certain alterations) from the Quart. J. Micr. Sci. XL. 1898, p. 613.

the posterior end, the whole body is hollowed so that each segment is curved. The most posterior segments, which are crowded with embryos well advanced in their development, are rounder, less flattened, longer, and they readily broke off.

I was not able to detect any genital pore on the exterior even with the aid of powerful lenses, but sections (figs. 4 and 6) and stained mounted specimens show that it is on the same side of the body in all the segments.

The head of the tapeworm bears four suckers, and in the midst of them is the rostellum (figs. 3, 8 and 9). The shape of the head is very various: in some cases the suckers are, as it were, hunched up and lie at each corner of a square, the lateral diameter of which does not exceed the dorso-ventral (fig. 8); in other specimens the head is not separated from the body by a deep constriction, but is flattened and spread out (fig. 7), so that the lateral suckers are separated from one another by a space considerably wider than that which lies between the dorsal and the ventral suckers.

The rostellum is minute and sunk in a pit (fig. 3); it bears a wreath of ten hooks. In all the specimens which I cut into sections, and I think in the others as well, the rostellum was retracted, the points of the hooks folded in against the axis of the rostellum, and not reaching so far forward as the mouth of the pit. When the animal is fixed to the mucous membrane of its host this rostellum is doubtless protruded from its sheath, and the hooks are divaricated. Certain muscle-fibres which run from the base of the rostellum, and lose themselves in the parenchyma, probably serve to retract it.

The hooks are slightly curved, and the projection which corresponds with the inner fork of the more triradiate hooks of other genera is hardly, if at all, marked (fig. 2). Measuring in a straight line from the base to the tip the hooks are $18-23\mu$ in length, thus corresponding pretty closely with those of *Drepanidotaenia tenuirostris* which, according to Raillict[1], measure 20 to 23μ, and to those of *D. lanceolata*, which measure 25 to 31μ.

The four suckers present no peculiarities; they are deeply cupped, with a small orifice to their lumen, but probably they are capable of considerable change of form (fig. 9). They are probably retracted by some muscle-fibres which cross one another and run into the parenchyma.

The segmentation of the body begins immediately behind the suckers: at first the segments are very short, but they gradually increase in size throughout the first three-quarters of the length of the body. For the last quarter the segments are crowded with embryos; they become in this region much narrower, more cylindrical in shape, and longer, and are very easily broken off. The posterior free edge of the segments of the anterior two-thirds of the body is sharp, and may overlap the segment behind, or may stand out clearly from it.

The water-vascular system is well developed; on each side of the body are two longitudinal canals,—one, the ventral, much bigger than the other, or dorsal. The

[1] Traité de Zoologie médicale et agricole, Paris, 1895.

lining of the former seems to be a structureless cuticle with no cells especially related to it, but the wall of the dorsal vessel is surrounded by a number of small deeply stained cells (fig. 4). I did not see any communication between the vessels of one side, but the larger vessels communicate as usual, one with another, by a transverse vessel running from side to side along the posterior border of each segment. In the head the vessels all communicate. In some of the better preserved sections such structures as are depicted in fig. 10 were seen; these may or may not be flame-cells; they look rather like them. No valves were seen in the course of the vessels.

The lateral nerve-cords are well marked, lying externally to the ventral excretory canals; they fuse together in the head, forming a ganglion which is indicated in fig. 3. No traces of the nerve-ring described by Tower[1] as running round the posterior end of each segment of *Moniezia*, or of the secondary nerves described by the same observer, were to be seen. But these, if present, probably require fresh material and special methods of preservation to make them manifest. Special nerve-cells, described below, are scattered through the parenchyma of the body.

The histology—at least in some specimens—could be fairly well made out, and agrees roughly with what Blochmann has described in *Ligula monogramma*[2]. The whole body is covered by a cuticle, the outer fifth of which stains more deeply than the remainder. Within this, with a high power, a number of dots or knobs become visible (fig. 10). These are the swollen terminations of certain strands or processes of the ectoderm cells. The cells themselves, as Blochmann has shown, lie removed to some distance from the cuticle they secrete, but are in contact with it by means of the above-mentioned processes ending in the knobs.

The ectoderm cells are not all at one level, but on the whole form a fairly well-marked layer. Each cell is fusiform in shape, and produced into two or three processes, which project both peripherally and centrally. They contain large and well-marked nuclei. Neither the cells nor their processes are laterally in contact; they are separated one from another to varying extents by the intrusion of some of the parenchymatous network which makes up so much of the body of a Cestode.

This parenchyma consists of a meshwork which permeates everywhere the body of the tapeworm, surrounding all the organs, and often, as is the case with the ectoderm and the muscles, passing in between their constituent cells. In the spaces of the meshwork there is believed to be a fluid. The meshwork itself is secreted and nourished by certain large star-shaped cells which are irregularly scattered through the parenchyma, and which give off processes in all directions (fig. 10).

Round the generative glands this parenchymatous network becomes condensed, the spaces disappear, and it forms a close sheath to the ovary, testis, &c. At the posterior end of each segment it is also somewhat condensed, and in section presents

[1] Zool. Anz. vol. xix. 1896, p. 323.
[2] Die Epithelfrage bei Cestoden und Trematoden, Hamburg, 1896.

the appearance of a well-marked double line, which is very characteristic, and is well shown in fig. 6.

Scattered amongst the parenchyma are certain faintly stained cells which seem to be bipolar, and which differ from the cells of the parenchyma both in shape and in their powers of absorbing the staining reagents. These I take to be nerve-cells which are in communication with the nerve-fibres of the lateral cords. The latter are entirely devoid of any nerve-cells on their course.

Muscle-fibres are scattered through the substance of the body, and one set of longitudinal muscles are most definitely arranged. This layer is situated just below the epidermis in the anterior part of the segment, but as the latter increases in size posteriorly, the cylinder of muscle-fibres, which retains the same diameter throughout, comes to lie more deeply in the tissues. These muscles, like the nervous system and excretory canals, run from segment to segment ; some of them, if not all, end in the cuticle, where it is most bent in at the posterior end of each segment. Laterally the fibres are not in contact, being separated by considerable intervals. Their regular arrangement is shown in fig. 5.

In the posterior segments, which are so ripe that the slightest touch breaks them off, the parenchyma has undergone considerable degeneration, the cells are less clear, and the spaces of the meshwork are larger and more irregular.

The generative organs begin to arise very early in the series of segments. Already in the eighth or tenth segment clusters of cells are segregating, and their deep staining shows that they belong to the gonads. In the sexually ripe segments the ovary is centrally placed, and is supported on each side by a lobe of the testis. From the latter a fine vas deferens leads into an extensive vesicula seminalis, which is as a rule crowded with spermatozoa ; from this a muscular duct leads to the unilateral genital pore. I was unable to make out the details of the penis, and similarly I failed to detect any yolk-gland amongst the female genitalia.

The vagina leads at once into a large receptaculum seminis, whose walls were strengthened by a series of cuticular-looking rings, whose cut ends are shown in figs. 4 and 6. This communicates both with the oviduct and with the uterus. The latter presents no special points of interest : in the posterior segments it contains the typical three-hooked larvae, each segment containing at least one hundred and probably more.

Classification.

In his paper on taenias in birds, Dr Fuhrmann[1] remarks that of the 240 odd species of tapeworm described from avian hosts, only twenty-one have been studied anatomically : the remainder are but little more than names, and probably many of the names are of doubtful validity.

[1] Rev. Suisse Zool. tome III. 1895—6, p. 433.

A certain amount of order has been introduced into this mass of material by the establishment of certain sub-groups, and by the giving of a new generic name to the members of these subdivisions; thus in 1891 Blanchard and Railliet[1] established the genus *Davainea*; in 1892 Railliet[2] suggested two new generic names, *Drepanidotaenia* and *Dicranotaenia*, for certain tapeworms inhabiting, for the most part, domestic birds. These are characterised chiefly by the nature of the hooks. In the following year Diamare[3] founded the genus *Cotugnia*, in which the generative organs are double and have two pores, but which is distinct from the genus *Dipylidium* of Leuckart. All these genera are characteristic avian tapeworms, and are, with but very few exceptions, confined to birds.

There is little doubt that the tapeworm which I have described above from the intestine of *Hemignathus procerus* corresponds with a *Drepanidotaenia* of Railliet[4], who defines his genus as follows:

"Tapeworms provided with a simple crown of uniform hooks, which are usually few in number; the outer limb (manche) of the forked base of the hooks is much longer than the inner (garde), which is always slight; the point is directed backwards when the rostrum is withdrawn. The majority live in the intestines of aquatic birds. Their larva is a Cysticercoid, and is found encysted in the bodies of small fresh-water Crustacea."

Railliet describes eight species of *Drepanidotaenia*; in one of these the genital pores are on alternate sides of the body in successive segments; the remaining seven species are unilateral in this respect, but they fall into two groups,—one, with three species, in which the number of hooks is eight; and the other, with four species, in which the number of hooks is ten.

It is to this latter group that we must add the tapeworm from *H. procerus*. The four species *D. anatina*, *D. sinuosa*, *D. setigera*, and *D. tenuirostris* differ *inter se* in several respects, but perhaps the simplest way of determining the species is by measuring their hooks. Of these four species, *D. hemignathi* most nearly resembles *D. tenuirostris*, which occurs in certain of the ducks; it differs, however, markedly in size, being when mature about $\frac{1}{5}$ to $\frac{1}{12}$ the length of the last named. It resembles *D. tenuirostris* in the length of its hooks in the head, which in the latter are $20-23\,\mu$, in the former are 18 to $23\,\mu$; but whereas the hooks of the embryo are about the same length in the new species, i.e. about $20\,\mu$, in *D. tenuirostris* they are but $7\,\mu$. The neck is short, not long as in the last-named species, and the eggs are small, about $40-50\,\mu$ in diameter, and spherical in shape, not cylindrical as Krabbe[5] figures them, with a length of $85\,\mu$. The hooks also differ in shape; those of *D. tenuirostris* have a much more strongly

---

[1] Mém. Soc. Zool. France, tome IV. 1891, p. 420.
[2] Ibid. tome XVII. 1892, p. 115.
[3] Boll. Soc. Napoli, ser. I, vol. VII. 1893. p. 9.
[4] Traité de Zoologie médicale et agricole, Paris, 1895. p. 298.
[5] Danske Selsk. Skr. VIII. 1870, p. 249.

developed process corresponding with the inner limb of the forked base than occurs in *D. hemignathi*.

The species, which I named after its host, may be characterised as follows :

(1)   *Drepanidotaenia hemignathi* Shipley.

D. hemignathi Shipley, Quart. J. Micr. Sci. XL. p. 620.

Length 1—2·2 centimetres ; breadth, in the middle of the body, 2 millimetres. Head flattened and compressed, rostrum with a crown of ten hooks ; each hook 18—23 μ in length, and with but a slight trace of the inner limb of the forked base. Neck short. The first segments are short, but they very soon (eighth or tenth) show traces of reproductive organs. Genital pore unilateral. The posterior limit of each segment is sharply defined, and forms an angle of about 45 degrees with the sides. Egg spherical, diameter about 40—50 μ. The three pairs of embryonic hooks measure about 20 μ each in length.

HAB.   *Hemignathus procerus*, Sandwich Islands : in the intestine.

(2) Mr Perkins has also given me two or three specimens of a tapeworm from a *Loxops*, sp. This bird, like the *Hemignathus*, is a member of the family Drepanididae, which is confined to the Sandwich Islands. Unfortunately the specimens are without their head, and I am unable to identify them. They differ markedly from the *Drepanidotaenia* described above.

(3)   *Echinococcus* ?.

Echinococcus is mentioned by Dr Lutz as occurring occasionally amongst cattle killed for the market.

(4)   *Taenia crassicollis* Rud.

This cysticercus larva of this species was found by Dr Lutz in *Mus decumanus*.

(5)   *Taenia solium* L.

The entozoon is said to be very uncommon, but is occasionally met with.

III.   ACANTHOCEPHALA.

(1)   *Apororhynchus hemignathi*¹ Shipley.

In the summer of 1894 I received from Mr Perkins seven small parasites which he had noticed adhering lightly to the skin around the anus, but beneath the skin, of a species of bird, *Hemignathus procerus*, which he collected in the island of Kauai. Each

¹ Quart. J. Micr. Sci. XXXIX. p. 207 and XLII. p. 361.

of these parasites was divided into three regions,—a head, a collar, and a trunk ; and, in fact, they have an almost ludicrous resemblance to a young *Balanoglossus* with one or two gill-slits (figs. 11, 12, and 13). On investigating their anatomy it at once became evident that the animals belonged to the group Acanthocephala, and, further, that they differed from the other members of the group in the absence of what is perhaps their most characteristic organ,—from which, indeed, they take their name—the hooked proboscis or introvert. Careful inspection failed to reveal any trace of a scar or mark where the introvert might have been broken off; and although in the absence of hooks and introvert sheath, &c., the anterior part of the body which I have called the head is as unlike the typical introvert as possible, still in its relation to the lemnisci and to the ligament it occupies the position of that organ, and until we can get further information I think the best plan is to regard this part of the body as equivalent to the eversible part of more normal forms.

The second of the three regions into which the body is externally divided is shorter than the head and smaller in diameter ; it may be termed the collar. The third or posterior region, which may be called the trunk, is the longest and the most slender of the three ; behind it tapers to a point where the orifice of the genital duct is situated, and this end of the animal is always a little turned up (figs. 11, 12, 13, 17 and 23). The exterior of the collar and trunk are smooth or lightly wrinkled, but the head is covered with a number of small depressions or pits which give it a very characteristic appearance, and which are well seen in sections. The head is attached to the collar by a narrow neck, which is surrounded and concealed by the edge of the collar. This is obvious in sections (figs. 15 and 23). All the specimens were somewhat shrivelled and apparently distorted. The largest measured 3·5 mm. in length, the smallest 2·5 mm. ; had they been fully distended they would probably have been 1 to 1·5 mm. longer. The body-cavity of the head is continuous with that of the neck, and the latter opens freely into the cavity of the trunk (fig. 23). The first-named space is by far the largest. The lumen of the collar region is reduced by the great thickness of the walls of this part of the body, and both here and in the trunk much of the internal space is occupied by the lemnisci and the reproductive organs.

The skin is one of the most characteristic features of the Acanthocephala, and as far as I know is only paralleled by that of the Nematodes, but it possesses certain features not found in the last-named group. The whole body is covered by a thin cuticle which does not vary much in thickness in the different regions of the body, and which is invaginated a short distance into the genital pore. Beneath this is the true epidermis, or subcuticle as it is called; this has in my specimens the usual structure met with in the group so well described by Hamann, and consists of a matrix of a fibrillar nature, the fibrils being as a rule arranged radially, in which are embedded a certain number of amoeboid nuclei (figs. 16 and 20). This tissue is much thicker in the region of the collar than elsewhere, and it is thicker in the trunk than in the head. It is pierced in all directions by a series of tubes or lacunae which have no definite lining, but which seem to be mere splits in the fibrillar matrix. The lacunae—except in the head—have a general circular direction which is very well marked in the trunk region where each runs into a lateral longitudinal split (figs. 20 and 24). They contain a small amount of coagulum, the remnant of the fluid which circulates in them; during life this fluid, in other species, holds in suspension fat and coloured oil globules. If these are present in my species they must have been dissolved out in the processes which precede embedding. The circular lacunae of the trunk not only communicate with one another by means of the two longitudinal lateral lacunae (figs. 20 and 24), but they open into one another by numerous small branches which have an oblique or longitudinal direction. In the head the lacunae have a general longitudinal course; they are not, however, straight, but twist in and out between the pits on the surface; they anastomose freely (fig. 14). Thus in a transverse section of the head the lacunae appear as round holes more or less uniformly arranged in the skin, and the same effect is produced by a longitudinal section of the trunk.

In the collar region the subcuticular tissue is much thickened, and the lacunar system forms a single more or less definite ring which gives off numerous branching anastomosing twigs (fig. 15).

Although the above account attempts to give the general course of the lacunae in the skin, it should be mentioned that there is considerable irregularity in the arrangement, and one is almost inclined to believe that the canals do not remain permanent, but that they sometimes close up and new ones appear. As they have no lining of any kind, such a closing would leave no trace.

As Schneider[1], Hamann[2], and Kaiser[3] have shown in the species investigated by them, the lacunar system of the introvert is completely shut off from that of the neck—if it be present—and of the trunk, by a fold inwards of the cuticle which cuts the subcuticular tissue in two. I have not been able to find any such cuticular ring in the species in question, but the state of preservation of my specimens does not allow me to say definitely that it does not exist.

The lemnisci are two elongated sac-like prolongations of the subcuticular tissue which are attached anteriorly to the skin at the junction of the head and collar. They extend backwards to the extreme posterior end of the body, and are slightly bent so that a longitudinal section may cut them in two or three places (fig. 23). Histologically they are composed of the same substance as the subcuticle in direct continuity with which they arise, and they are traversed by a similar system of canals. Physiologically they seem, as Hamann suggests, to act as reservoirs for the fluid of the canal system of the introvert; when the fluid they contain is forced into the spaces of the introvert the latter is everted. It is withdrawn again into the body by special muscles. In most

[1] Arch. Anat. 1868, p. 584.
[2] Die Nemathelminthen, Heft 1 and 2, Jena, 1891 and 1895.
[3] Bibl. Zool. Heft 7, 1892, p. 1.

species the canal system of the lemnisci opens into that of the introvert in front of the cuticular ring, and is thus completely independent of that of the trunk. If we assume that the head of my species corresponds with the introvert of other forms which have lost its introvert sheath, the lemnisci open into the same region of the skin as they do in other Acanthocephala.

The nuclei of the subcuticle and of the lemnisci are very remarkable; they correspond in structure with those described by Hamann in *Neorhynchus clavaceps*, in which species according to this observer both the skin and the lemnisci retain in the adult their embryonic condition. As in *Neorhynchus* the number of nuclei is very small, some twelve to twenty seem to suffice for the whole of the subcuticle, and perhaps two to four for each lemniscus. The structure of the nucleus shows a most striking resemblance to an amoeba with rather short pseudopodia (figs. 16, 20, and 23). No single nucleolus can be detected, but numerous chromatin particles are present, and in some a distinct vacuole can be observed. These nuclei are scattered about in a most irregular fashion; not one may be seen in a number of consecutive sections, and then perhaps three or four may appear, and from their large size persist through several sections. The nuclei lie, as a rule, embedded in the substance of the subcuticle; more rarely they are found in the lacunae. Although there is no proof, one is tempted to believe that the nuclei wander through the subcuticle and lemnisci in an amoeboid manner, and that the small number of nuclei which are found in these tissues is compensated for partly by the large size of each, but more especially by their mobility. Similar amoeboid nuclei undoubtedly move about, fuse with one another, and undergo fission in the subcuticle of the larval forms of *Neorhynchus clavaceps*.

Within the subcuticle and completing the skin on the inner side, is a layer of circular muscles, and still more internally a layer of longitudinal muscles (figs. 16 and 25). The muscles of these layers are but a single fibre thick, and they are not very uniformly present. The circular layer is most complete in the region of the trunk, and I have figured a section to show this (fig. 22). The longitudinal layer is even less definite, but scattered fibres can be detected here and there (figs. 16 and 25). Each fibre appears to be spindle-shaped, and in the circular muscles has the striated portion only on its outer face, forming a thin band; the inner half of the fibre consists of vacuolated strands of protoplasm in which is a nucleus. The longitudinal layer of muscles alone is continued over the lemnisci (figs. 19 and 24). These muscles are not covered on their inner side by any layer of epithelial cells, neither does any such layer cover the ligament, but both tissues lie freely exposed to the fluid of the body-cavity.

In the more typical Acanthocephala the anterior end of the body terminates in a hollow eversible portion provided with rows of hooks whose number and shape have a certain systematic value. This introvert can be withdrawn, not into the general body-cavity, but into the cavity of the introvert sheath, which is shut off from the general body-cavity by a double (Echinorhynchidae) or a single (Neorhynchidae) wall. The extrusion of the introvert is believed to be effected by fluid being forced into its lacunae by the lemnisci. It is retracted by special muscles attached to the inside of its tip; besides these, other retractor muscles run from the outside of the introvert sheath, and these serve to retract the whole sheath and its contents into the trunk. The chief nerve ganglion lies as a rule on the posterior end of the introvert sheath, usually in the middle line, but in the Gigantorhynchidae it is placed to one side. From the posterior end of the introvert sheath, and having its origin between its two walls when they are present, the ligament runs backward, traversing the body cavity, and ending in the funnel-shaped internal opening of the oviduct in the female and in the vas deferens in the male.

Owing to the absence of an introvert and its sheath, the relations of the ligament in the present species is somewhat altered. It takes its origin from the anterior end of the head, and at first seems to consist of a few strands of muscular fibres which arise from the muscles of the skin (fig. 21). All my specimens but one proved to be mature females, whose ovaries had broken up into the egg masses which are characteristic of the Acanthocephala. These egg masses consist of packets of a dozen or more cells of which the peripheral layer develop into ova at the cost of the central cells which serve them as a food supply (figs. 14, 16, and 23). These packets coexisted in my specimens with ova in various stages of development, some without any egg shell, whilst others were provided with a thick deeply-staining membrane. The whole lumen of the head was crowded with these ova. In the region of the collar the ova were confined by a thin-walled membrane, and in the trunk there were two such masses of ova, which, however, seemed less mature than those lying in the head. Lying amongst the various organs in the body-cavity were a number of very finely granular masses, which I take to be the masses of spermatozoa (figs. 16 and 20). Of the complex system by means of which the ova leave the body, little could be made out beyond the fact that a well-marked funnel is present opening into the posterior end of the body-cavity of the trunk (fig. 19). I failed, however, to find a second opening near the narrow end of the funnel such as occurs in other forms, but this may have been due to the poor state of preservation. The funnel leads into a duct which opens on the posterior end of the trunk.

The testes are two in number, and lie one behind the other in the ligament, though owing to its looping both may appear in the same transverse section. The spermatozoa do not escape into the body of the male as the ova do into that of the female, but pass down a duct in the ligament which opens at the end of the body. Traces of accessory glands were seen, but the details were not clear.

The brain lies on or in the ligament just behind its point of attachment to the skin of the head (figs. 21 and 23). Owing to the disruption of the ovaries in my female specimens the ligament could not be traced very far, but in the only male it reached from one end of the body to the other. The brain consists of a few large ganglion cells with a clear homogeneous cytoplasm and deeply-stained nuclei; the divisions between the cells were very sharp and straight (fig. 21). In the females this mass of cells lay on the ligament: in the male, on the other hand, it occupied the centre of the fibrous and muscular strands which compose that body (fig. 25). In the former I could trace no nerves leaving the brain, but in the male two nerves surrounded by muscles pass backward; these obviously correspond with the retinacula of other forms.

Classification.

Until recently the group Acanthocephala included but one genus, *Echinorhynchus*, which comprised several hundred species. Recently, however, Hamann[1] has pointed out that these species present certain differences which enable him to divide the group into three families, each with a corresponding genus. To these I venture to add a fourth, to include the remarkable form above described. This family may, I think, be called the Apororhynchidae, and the new genus *Apororhynchus*[2], which name refers to the absence of the eversible introvert; and, inasmuch as it is convenient in naming a parasite to have some indication of its host, I think the specific name may be *hemignathi*.

If these terms be adopted, the classification of the Acanthocephala will be as follows, the characteristics of each of the first three families being taken from Hamann's papers.

1. Family ECHINORHYNCHIDAE. The body is elongated and smooth. The introvert sheath has double walls, and the introvert is invaginated into it. The nerve ganglion is in the introvert sheath, mostly embedded in it and central in position. The hook papillae are only covered with chitin at their apex, and the hooks have a process below.

Genus *Echinorhynchus*, with the characters of the family.

The vast majority of Acanthocephala belong to this family; a few may be mentioned. *E. proteus*, found in many fishes and varying in size with its host; its larval forms inhabit the Amphipod *Gammarus pulex*, and are also found in the body-cavity of numerous fresh-water fishes. *E. clavula* occurs in many fishes and in the intestine of a species of *Bufo*. *E. angustatus* is found also in fishes, with its larval form in the Isopod *Asellus aquaticus*. *E. moniliformis* is said to attain maturity in the human intestine; its usual host is a mouse, and its larval host is the larva of a beetle, *Blaps mucronata*. *E. porrigens* invests the intestine of the rorqual, and *E. strumosus* that of a seal. There are many others.

---

[1] Loc. cit. and Zool. Anz. Bd. xv. 1892, p. 195.

[2] In my original paper I suggested the name *Arhynchus*, but as Professor C. Wardell Stiles and Professor A. Hassall have pointed out that this name is preoccupied, having been used by Dejean in 1834 for a beetle, I later (Quart. J. Micr. Sci. XLII. p. 361) suggested the name *Apororhynchus*.

II. Family GIGANTORHYNCHIDAE. Large forms, whose body is ringed and flattened during life like that of a *Taenia*. The hooks are like those of a Taenia, the hook-papilla being entirely covered with chitin. There are two root-like processes in each hook. The introvert is muscular, has no lumen, and the introvert cannot be retracted into it. but the whole retracts into the body-cavity. The ganglion is excentrically placed to the side, behind the middle of the so-called sheath. The body-cavity is enclosed in a structureless membrane, and is traversed by membranes stretched transversely. The lemnisci are long. coiled, with a central lacuna.

Genus *Gigantorhynchus*, with the characters of the family.

Hamann includes three species in this family—*G. echinodiscus, G. taenioides*, and *G. spira*; and points out that *E. gigas* agrees with them in all points but that of the external annulation. The first of the above-named species occurs in the intestine of anteaters, and has been found in *Myrmecophaga jubata* and *Cycloturus didactyla*. *G. taenioides* has been found in a species of *Cariama, Dicholophus cristatus*; and *G. spira* lives in the king vulture. *Sarcorhampus papa. E. gigas* in the adult stage occurs in the small intestine of swine, and its larval host is believed to be the grubs of *Melolontha vulgaris* and *Cetonia aurata* in Europe and of *Lachnosterna arcuata* in the United States[1]. It is recorded once from the human intestine.

III. Family NEORHYNCHIDAE. Sexual maturity is reached in the larval state. The introvert sheath has a single wall. A few giant nuclei only are found in the subcuticle and in the lemnisci. The circular muscles are very simply developed, and the longitudinal muscles only present in places.

Genus *Neorhynchus*, with the characters of the family.

This genus includes but two species, *N. clavaeceps* and *N. agilis*. They both present interesting cases of paedogenesis, the large embryonic nuclei of the young larva do not break up into numerous nuclei as they do in the commoner species. *N. agilis* is found in *Mugil auratus* and *M. cephalus*; *N. clavaeceps* in the Carp, *Cyprinus carpio*, its larva form according to Villot[2] in the fat bodies of the Neuropterous insect *Sialis niger*; it has also been found in the alimentary canal of the leech *Nephelis octocula*, and specimens of the water-snail *Limnaea* have been artificially infected with it.

IV. Family APORORHYNCHIDAE. Short forms, with the body divided into three well-marked regions.—head, collar, and trunk. The head is pitted, the collar smooth, and the trunk wrinkled, not annulated—in spirit specimens. There is no eversible introvert, and no introvert sheath, and no hooks. The sub-cuticle and the lemnisci have a few giant nuclei, and the lemnisci are long and coiled.

[1] C. W. Styles, Zool. Anz. xv. 1892, p. 52.
[2] Zool. Anz. viii. 1885, p. 19.

Genus *Apororhynchus*, with the characters of the family.

This family in the length and curvature of its lemnisci resembles the Gigantorhynchidae, and in the persistence of the embryonic condition of the nuclei in the sub-cuticle and the lemnisci, the Neorhynchidae; but in the shape of the body, its division into three well-marked regions, the absence of eversible introvert, introvert sheath, and hooks, it stands alone, though to some extent nearer to the Neorhynchidae, in which the introvert is relatively small, the introvert sheath simple, and the number of hooks reduced, than to either of the other families.

The single species *Apororhynchus hemignathi* was found attached to the inner side of the skin in the neighbourhood of the anus of a Sandwich Island bird, *Hemignathus procerus*. This bird is a member of a family Drepanididae, which is entirely confined to the Sandwich Island group. Professor Newton tells me that it is probable that the "food of *Hemignathus* consists entirely of insects which it finds in or under the bark of trees"; hence it is probable that the second host of this parasite, if such exists, must be looked for amongst the Insecta.

(2) *Echinorhynchus campanulatus* Dies.

Found by Dr Lutz in Water-rats. This species is said to be a facultative parasite of man.

THE ZOOLOGICAL LABORATORY, CAMBRIDGE.

March, 1900.

DESCRIPTION OF PLATE XI. (VOL. II.)

MOLLUSCA.

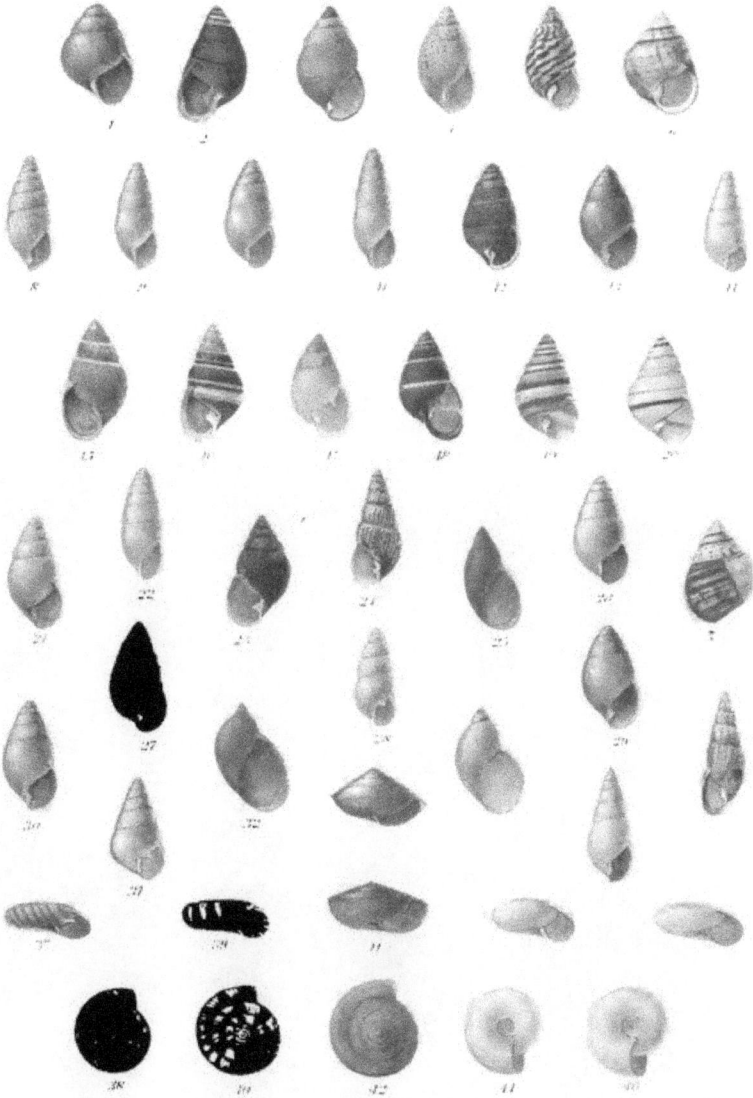

DESCRIPTION OF PLATE XII. (VOL. II.)

MOLLUSCA.

PHILONESIA BALDWINI Ancey (p. 281).

Fig. 1.   Mantle margin, with shell and dorsal lobes.   *a*, denotes white spots on integument covering the branchial cavity.
Fig. 2.   Extremity of foot from left side.
Fig. 2 *a*. Portion of foot showing sole.
Fig. 3.   Genitalia, not quite complete.
Fig. 3 *a*. Male organs enlarged, to show the retractor muscle and coiled vas deferens.
Fig. 4.   Jaw.
Fig. 5.   Centre and two side teeth of radula.
Fig. 5 *a*. Median teeth, 7th—11th.
Fig. 5 *b*. Lateral teeth, 19th—22nd.
Fig. 5 *c*. Eleven of the outermost, or marginal, teeth.

GODWINIA CAPERATA Gould (p. 277).

Fig. 6.   Animal with shell removed, showing dorsal lobes.
Fig. 7.   Portion of mantle zone, near respiratory orifice.
Fig. 8.   Extremity of foot.
Fig. 9.   Buccal mass, and salivary gland, &c.
Fig. 10.   Sole of foot.
Fig. 11.   Jaw.
Fig. 12.   Central tooth of radula.
Fig. 12 *a*. Fifth intermediate, and following lateral teeth.
Fig. 12 *b*. Outermost, or marginal, teeth.

Fig. 13.   *Leptachatina acuminata* Gould.   Central teeth of radula (p. 357).
Fig. 13 *a*. Intermediate and marginal teeth.
Figs. 14, 14 *a*. *Ancylus sharpi*, sp. nov. (p. 394).

EXPLANATION OF THE LETTERING.

Al. Gd.   albumen gland.
B. m.   muscle of buccal mass.
Gen. ap.   generative aperture.
h. d.   hermaphrodite duct.
i.   intestine.
l. d. l.   left dorsal lobe.
l. s. l.   left shell lobe.
P.   male organ.

pr.   prostate.
r. d. l.   right dorsal lobe.
r. m. P.   retractor muscle of penis.
r. s. l.   right shell lobe.
Sal. gld.   salivary gland.
st.   stomach.
v. d.   vas deferens.

The numerical digits in smaller type indicate the tooth figured, reckoning from 0, the central tooth.

All figures (except 14, 14 *a*) are from dissections and drawings made by Lt.-Col. H. H. Godwin-Austen, F.R.S.

DESCRIPTION OF PLATE XIII. (VOL. II.)

ENTOZOA.

LIST OF ABBREVIATIONS.

b.	Brain.	*n. s.*	Central nerve ganglion.
c. m.	Circular muscles.	*n.*	The amoeboid nuclei of the skin and the
c.	Cuticle.		lemnisci.
d. e. c.	Dorsal excretory canal.	*o.*	Ovary.
e.	Ectoderm.	*p.*	Nucleus of parenchymatous cell.
e. m.	Masses of ova.	*p. c.*	Parenchyma cell.
f. c.	? Flame-cell.	*p. e.*	Knob-like ends of ectoderm cells under
g. d.	Genital duct.		cuticle.
g. p.	The external opening of the duct.	*r. s.*	Receptaculum seminis.
l. & la.	The lacunae in the skin.	*r.*	Rostellum.
le.	The lemnisci.	*s.*	Coagulated masses of spermatozoa in the
li.	The ligament.		body-cavity of the female.
l. l.	The large lateral lacunae of the trunk.	*t.*	Testis.
l. m.	Longitudinal muscles.	*u.*	Uterus.
l. n.	Lateral nerve.	*v. e. c.*	Ventral excretory canal.
m.	The muscles from which the ligament arises.	*v. s.*	Vesicula seminalis.
n. c.	Nerve cell.		

Fig. 1. A view of *Drepanidotaenia hemignathi*, × 15. The dark patches in the anterior two-thirds of the body are caused by the generative organs ; in the posterior third they represent the eggs in the uterus.

Fig. 2. An isolated hook from the rostellum, × 500.

Fig. 3. A longitudinal section through the head, × 100. The rostellum, *r.*, is retracted. The point of fusion of the two lateral nerves is shewn at *n. s.* The section passes between the suckers.

Fig. 4. A transverse section through a mature proglottis, × 70.

Fig. 5. A longitudinal section, somewhat oblique, showing the regular arrangement of the longitudinal muscles, × 50.

Fig. 6. A longitudinal section through several mature proglottides, × 50. This shows the transverse connection between the two ventral longitudinal excretory canals and the transverse lines formed by the concentration of the parenchyma at the posterior end of each proglottis.

Fig. 7. A view of the head in an expanded, flattened-out state, × 60.

Fig. 8. A view of another head in a contracted, bunched-up condition, × 40.

Fig. 9. A transverse section through the head, showing the ten hooks on the rostellum and the four suckers.

Fig. 10. A portion of a proglottis, highly magnified to show the minute anatomy, × 450.

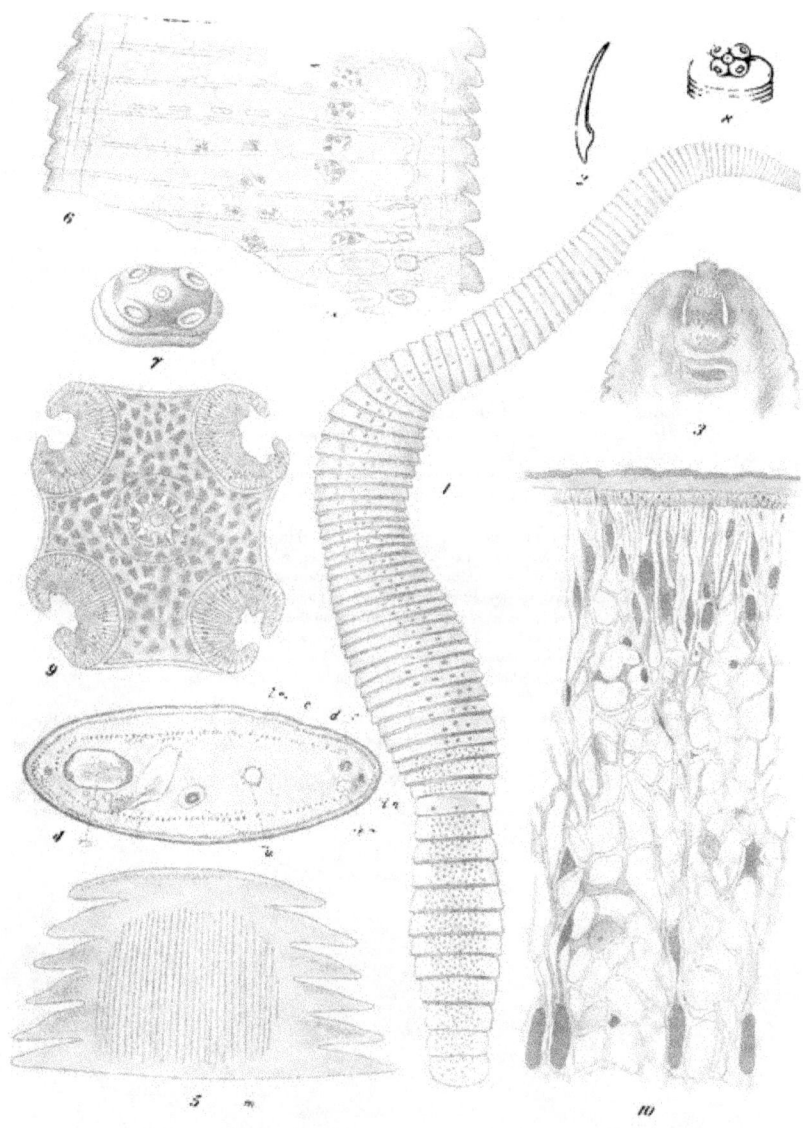

DESCRIPTION OF PLATE XIV. (VOL. II.)

ENTOZOA.

For explanation of letters see description of Plate XIII.

Figs. 11, 12, and 13. Three views of three different specimens of *Arhynchus hemignathi*. Each × 20.
The division of the body into three regions is well marked. The details are shown in Fig. 11.
Figs. 12 and 13 are rough sketches.

Fig. 14. A transverse section through the head of a female, crowded with ova and egg-masses; the
ligament is shown in section, × 40.

Fig. 15. A transverse section through the same, just below the edge of the collar. In the centre is the
neck, which fuses with the collar a few sections further back. The big circular canal of the
collar is shown at *l.*, × 40.

Fig. 16. A transverse section through the trunk of the same. The uppermost lemniscus is cut in two
places. The ovary is double, and shows egg-masses as well as eggs; some coagulated
masses of spermatozoa are lying in the body-cavity, × 40.

Fig. 17. A surface view of the external opening of the genital duct, × 40.

Fig. 18. Some developing ova, highly magnified.

Fig. 19. A transverse section through the trunk near the genital pore, taken from the same series as
Figs. 14, 15, and 16. It shows part of the funnel-shaped internal opening of the genital
duct, *g. d.*, × 40.

Fig. 20. A transverse section from another specimen taken behind the opening of the genital duct.
This shows the arrangement of the lacunae and their communications with the lateral
lacunae, *l. l.*

Fig. 21. A longitudinal section through the central part of the skin of the head, showing the origin of
the ligament and the ganglion cells of the brain, lying in a mass of ova and egg-masses.

Fig. 22. A small portion of the skin in section, showing the single layer of circular muscle-fibres, × 40.

Fig. 23. A median longitudinal section through a female. The whole body-cavity full of ova and
egg-masses. The ligament is seen in the head, and the genital duct near its opening in the
trunk. The left lemniscus, cut twice, is alone seen, × 30.

Fig. 24. A transverse section through the trunk of a male, showing one of the testes. This section
shows also the longitudinal muscles on the lemnisci and the large lateral lacunae, *lat. lac.*,
× 40.

Fig. 25. A transverse section through the head of a male, showing the brain in the ligament, and the
longitudinal muscle-fibres very well developed, × 40.

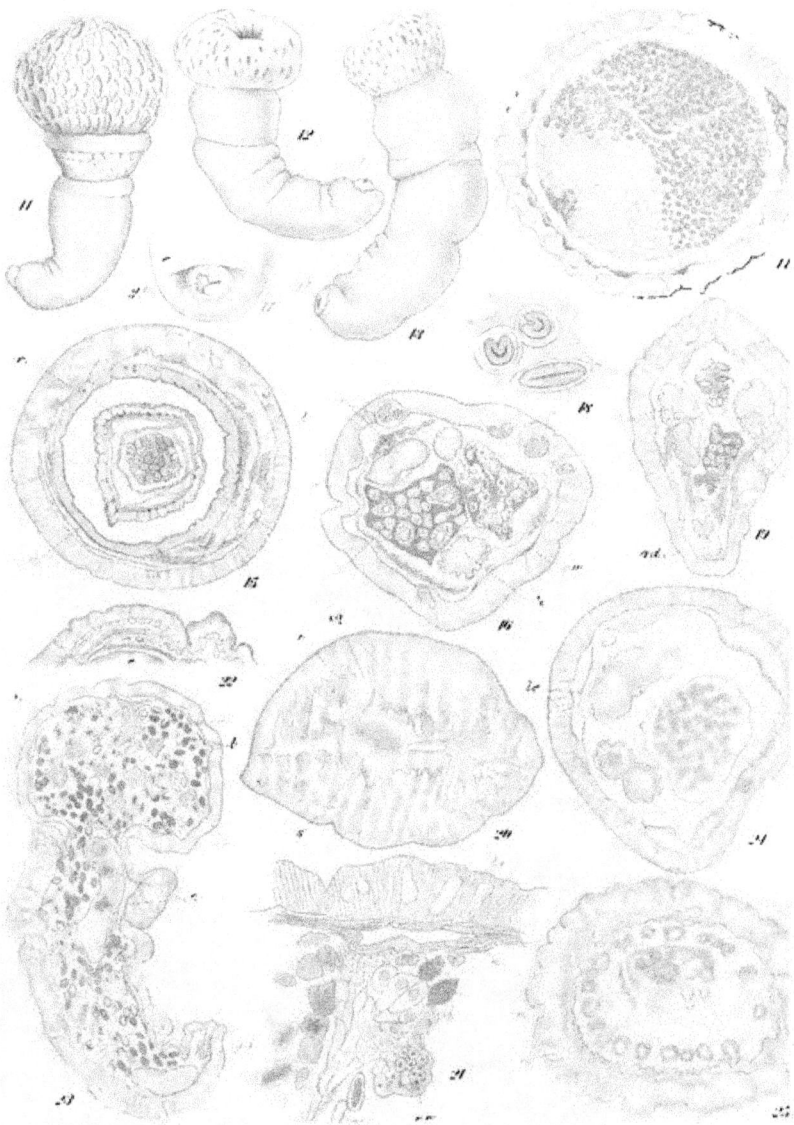

www.ingramcontent.com/pod-product-compliance
Lightning Source LLC
Chambersburg PA
CBHW021802190326
41518CB00007B/410